Electric Circuit Theory

Other Titles in the Series

A Computer Illustrated Text

ELECTRIC CIRCUIT THEORY

B E Riches

Department of Electronic and Electrical Engineering, Leicester Polytechnic

Adam Hilger, Bristol and Philadelphia
ESM, Cambridge

British Library Cataloguing in Publication Data

Riches B. E.
  Electric circuit theory
  1. Electric equipment. Circuits. Theories
  I. Title II. Series
  621.319′2′01

ISBN 0-85274-041-7 (pbk.)
ISBN 0-85274-043-3 (IBM disc)
ISBN 0-85274-044-1 (BBC 40/80 track disc)
ISBN 0-85274-042-5 (Network pack)

Library of Congress Cataloging-in-Publication Data

Riches, B. E.
  Electric circuit theory.
  (A Computer illustrated text)
  Includes index.
  1. Electric circuits. 2. Electric currents.
I. Title. II. Series.
TK454.R515 1989   621.319′2   88-34748

Series Editor: **R D Harding**, University of Cambridge

Published by: IOP Publishing Ltd
                 Techno House, Redcliffe Way, Bristol BS1 6NX,
                 England
                 242 Cherry Street, Philadelphia, PA 19106, USA
                 ESM, Duke Street, Wisbech, Cambs PE13 2AE,
                 England
Published under the Adam Hilger/ESM imprint

Printed in Great Britain by J W Arrowsmith Ltd, Bristol

To my daughter Gillian

To my daughter Gillian

⟩ Contents

〉 Preface

Electric circuit theory is one of a range of subjects where the presentation can be enhanced by adopting the format of a computer illustrated text (CIT).

In this instance the integrated software is used to illustrate the text in essentially two ways. The first produces dynamic and flexible illustrations and the second provides integrated problem-solving programs for demonstrating the theory and for subsequent analysis. The programs are identified throughout the text where they would normally first be used. The problem-solving software is additionally intended to be used as required.

The treatment of electric circuit theory given is introductory. For some students it will cover the subject to an appropriate standard. For others it is hoped that the CIT format will strengthen their understanding of fundamental principles, while the emphasis on computer methods forms an appropriate foundation from which to develop a competence in the use of professional electronic computer-aided design (ECAD) tools.

The approach used does not rely on advanced mathematics and is suitable for first year degree and diploma students. The emphasis on the appreciation of circuit behaviour, by the use of graphical responses, makes this CIT particularly suitable for students wishing to combine an understanding of electric circuit theory with other disciplines, as well as those specializing in electronic engineering.

The CIT has particular qualities as an educational aid. The format is well suited to open learning in the study or laboratory. In the latter case, the CIT can be used in computer integrated practical assignments with the software generating supportive data for laboratory demonstrations of circuit behaviour. For the lecturer the dynamically developed illustrations and function driven graphs should prove to be a useful source of visual aids.

A selection of practice problems is given at the end of each chapter. The problems are an integral part of this CIT and in some instances extend the theory, as well as providing practice in using the software for problem solving. Answers to the practice problems are included in the software. Appendices are used for developments beyond the style of the main text, general information on the software and a statement on units.

Finally, on a personal note, I would like to acknowledge the help and encouragement I received from Ray and Noreen Butler in the early stages of preparing this CIT and the useful comments made on the manuscript by John McKay.

B E Riches
Leicester
1988

〉 Chapter 1

〉 Introductory Concepts and Conventions

〉 1.1 Overview

The objectives of the first chapter of this book are to some extent different from the remainder and an overview of the strategy adopted is given below.

Students of any subject approach it from a variety of different educational backgrounds and one objective of this chapter is to present the necessary background material required for the later ones. The standard units of measurement for the various circuit theory parameters are stated and related to each other as they are introduced; a more precise definition of the units and other related information is included for reference in Appendix C. There is some choice regarding conventions and an attempt has been made to include the more common ones to form the basis for the subsequent analysis.

Chapter 1 is also introductory and representative of the illustrative and problem-solving programs used. The remainder of the software has similar aims but performs more sophisticated tasks.

Electronic engineering and electric circuit theory
Electronic engineering is fundamentally concerned with the controlled separation and movement of electric charges. The processes involved are complex and varied and produce a wide and continually developing range of effects which are utilized in an ever increasing range of products.

Electric circuit theory is concerned with modelling physical systems of moving charges, thereby contributing to basic understanding and providing the mechanisms for simulation, analysis and synthesis. It therefore plays an important part in the various levels of electronic engineering practice which result in product design and development.

Systems and models

The term circuit is applied widely to describe an electronic system, e.g. an amplifier, a model of a system and a symbolic representation of a model in the form of a circuit diagram.

Electric circuit theory is concerned with the rules which govern the circuits which are used to model electronic systems.

The relationship between such systems and the models used to represent them can be illustrated by the processes of analysis and synthesis.

Analysis may be performed on the model of a system in order to interpret the system behaviour. The requirements of the model are that it is simple enough to analyse and realistic enough for the results to be meaningful in relation to the system. Synthesis, on the other hand, produces a model which behaves in a manner which corresponds to that required of a system. A working system can then be produced from the model, provided it is realizable and again realistic enough for its behaviour to predict that of the system. Thus although circuit models are completely governed by a set of theoretical and precisely defined rules, for the models to be effective in engineering terms there must be a close two-way relationship of representation and realization between model and system.

Modelling process

The modelling process is illustrated in broad terms by the diagram in figure 1.1.

For representation the procedure is as follows.

  (i)   A process model is formed of the system;
 (ii)   a circuit model is formed of the process model; and
(iii)   a circuit diagram is drawn which represents the circuit model and the system.

Stages (i) and (ii) may be omitted once the procedure is well understood.

The procedure for realization is the reverse of that for representation.

Simplifications, and therefore inherent approximations, occur in forming the process and circuit models. The elements in the circuit model obey precisely defined rules and the circuit diagram is an exact symbolic representation of the circuit model.

The process model may vary from little more than a simple mental picture of how a system operates to a representation of state-of-the-art physical science. This will depend on the precision required of the model.

Sections 1.2–1.5 describe the basic circuit theory models and relate them to the physical processes they represent. Charge, current and voltage are first described and then the properties of the circuit elements are defined in terms of voltage and current. The remainder of the book then uses these models in various combinations relating them to actual systems as appropriate.

The intention is to keep the process models as simple as possible, compatible with an appropriate appreciation of the system being modelled.

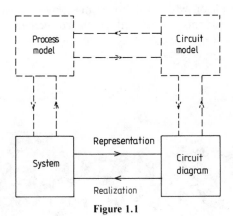

Figure 1.1

〉 1.2   Charge, current and Kirchhoff's current law

1.2.1   Charge

Positive and negative charges of electricity are naturally occurring constituents of the material world. Both types of charge occur in equal quantities making matter as a whole electrically neutral. This means that charge separation is a precursor of its control. The smallest quantity of electric charge exists on an atomic scale and is the charge carried by a single proton or electron. These elementary charges are equal to $\pm 1.6 \times 10^{-19}$ coulomb respectively. The process model of positive and negative charges treats them as continuously variable quantities of electricity.

The unit of electric charge is the coulomb (C).

1.2.2 Current

Electronic circuits and systems consist of conducting paths, along which electric charges are carried through a variety of materials and other physical circumstances. Some common examples are wires, strips on printed circuit boards, inside VDU tubes, electrolytes in batteries, inside integrated circuits, in resistors and under the influence of electromagnetic fields in capacitors and inductors. A continuous conducting path constitutes a closed circuit.

Figure 1.2 (reproduced by permission of Harrison Information Technology Ltd) shows an electronic circuit consisting of a variety of components mounted on a printed circuit board. The photograph also shows the heat sink on which the board and power devices used in the system are mounted; this is provided to absorb the heat associated with the conduction process.

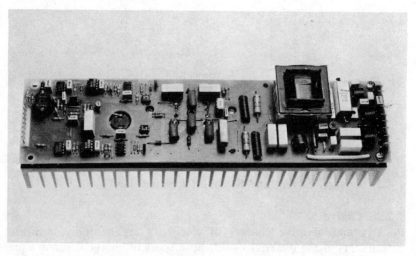

Figure 1.2

Movements of charge carriers through the various parts of electronic circuits are complex physical processes. The process model of current used here is by contrast relatively simple, considering currents along circuit paths to consist of negative, positive, or negative and positive, charge flow.

The moving charge may be part of the total constituent charge of a material (conduction current), or it may be moving through a vacuum (convection current). These examples may model the charge flow in a copper wire and through a VDU tube, respectively. There are important physical differences between conduction and convection currents but circuit theory does not differentiate between them.

Program filename 'CURRENT' produces an animated version of the model. The positive and negative charges of electricity shown in figure 1.4 represent only those associated with the current and not the total charge within the material.

1.2.3 Convention

For circuit analysis the current magnitude and direction must be specified and this is done by attaching arrows with accompanying labels to circuit diagrams. The arrow direction and the quantity ascribed to it must be interpreted together in order to specify the current completely. The convention adopted is demonstrated by reference to figures 1.3 and 1.4.

$$I \qquad\qquad\quad = \qquad\qquad -I$$

Figure 1.3

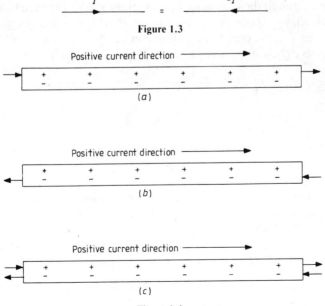

(a)

(b)

(c)

Figure 1.4

The interpretation of the convention is that the current in the direction of the arrow is equal to the ascribed quantity. The relationship between the sign of the ascribed quantity and the current direction is illustrated by figure 1.3. For equality a reversal of sign must accompany a reversal of direction: for example, if $I = -1$ ampere, then the current is 1 ampere in the direction right to left.

Figure 1.4 relates the sign and direction of the charge flow to the corresponding current. If the current is produced by a positive charge flow, the direction of positive current is the same as that in which the charges are moving. If the current is produced by moving negative charges then the direction of positive current is opposite to that in which the charges are moving.

Positive current corresponds to the term conventional current and current used on its own in some contexts may refer to positive or conventional current.

The unit of electric current is the $C\ s^{-1}$ or ampere (A).

1.2.4 Current and charge calculation

Program filename 'I' demonstrates the principle of current calculation in terms of charge flow. The positive and negative charges shown in figure 1.5 move through the conductor and the current is determined by the rate at which charge Q passes the fixed point P. The sign and magnitude of the current I is computed in the direction of the charge movement. This is equal to dQ/dt, the signs of Q and I being the same.

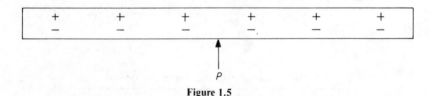

Figure 1.5

Suggested examples for the program are to determine the current when each moving charge represents: (i) one electron; (ii) one coulomb, and (iii) 6×10^{22} electrons.

Since current is the rate of charge flow it follows that the charge Q moved by an average current I_{av} A in t s is given by

$$Q = I_{av}t\ C$$

in the direction of the current.

1.2.5 Current continuity, Kirchhoff's current law

If there is no build-up of charge at a circuit junction the current is continuous. In other words, the current flowing away from the junction is equal to the current flowing towards it. This property of electric current is usually referred to as Kirchhoff's current law; it is illustrated for a circuit junction in figures 1.6 and 1.7.

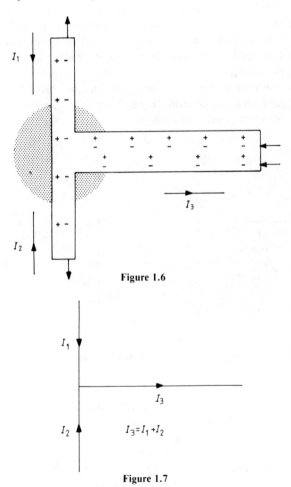

Figure 1.6

$$I_3 = I_1 + I_2$$

Figure 1.7

Figure 1.6 represents a junction in part of an electric circuit, enclosed by an imaginary surface. Within the closed surface the resultant charge is always zero. This means that whatever charge flows into the enclosed volume, where part of the surface intersects a circuit conductor, it flows out again at another surface–conductor intersection.

Program filename 'KCL' animates figure 1.6 to demonstrate this property and hence current continuity.

Alternative expressions
Using the sign of the current variable in relation to the indicated direction produces the two alternative expressions of KCL shown in figure 1.8. Thus if *all* the currents flowing at a circuit junction are shown directed (i) towards, or (ii) away from, the junction, then in either case the sum of the currents is zero.

For the animation produced by program filename 'KCL', I_1, I_2 and I_3 are all positive, but in general currents may be positive or negative relative to their indicated direction.

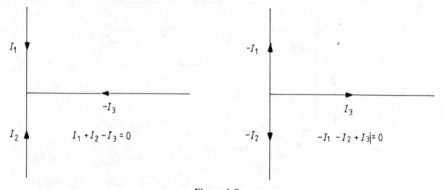

Figure 1.8

⟩ 1.3 Potential difference and Kirchhoff's voltage law

1.3.1 Potential difference
As electric charges move around a system they gain or lose energy. This transfer of energy between the moving charges and the system is modelled and quantified in terms of electrical potential. Positive charge gains energy in moving from a point of lower to a point of higher potential and loses energy in moving from a higher to a lower potential. The opposite applies to negative charge which gains energy in moving from higher to lower potentials.

Figure 1.9 shows a voltage source and a resistance, indicating the high- and low-potential terminals, the current *I* being positive in the direction shown. These are common examples of an energy source in which charge gains energy and an energy sink in which it loses it.

Energy storage elements such as capacitance and inductance can operate as both sources and sinks, depending on whether the electromagnetic fields associated with them are increasing or decreasing in strength.

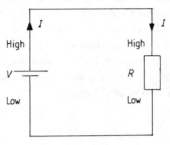

Figure 1.9

Potential is quantified in terms of unit charge and the difference in potential between two points in a circuit is equal to the energy gained or lost by each coulomb of charge transferred from one point to the other. Potential difference is therefore the property of two points in a circuit. The energy associated with the potential difference is that acquired or lost by charge moved between the two specified points.

Thus if a charge of Q C gains or loses W joule of energy in moving between two points in an electric circuit then a potential difference given by

$$V = W/Q \text{ volt}$$

exists between the two points.

The unit of energy is the joule (J). The unit of potential difference is the J C^{-1} or volt (V).

1.3.2 Conventions

There are a number of alternative terms and conventions for specifying potential differences. Voltage, voltage rise and voltage drop are alternative expressions for potential difference and three commonly used methods of labelling circuit diagrams are shown in figure 1.10.

(*a*) V_{ab} denotes the potential of point a with respect to, or relative to, point b. V_{ab} is positive when a is higher in potential than b and negative when it is lower.

(*b*) The arrow is used as a pointer. The quantity attached to the arrow is the potential of the end of the element to which it is pointing relative to the other end. For a voltage source the arrow indicates the direction of current flow through the source at any instant when the quantity attached to the arrow is positive.

(*c*) The quantity attached to the circuit element is the potential of the end labelled ' + ' relative to the other end. At any instant when the quantity attached to the circuit element is positive the polarity of the voltage across the element is as shown.

Linking the conventions gives $V_{ab} = V$ and $V_{ba} = -V$.

The approach adopted in this text and software is to use voltage for the general expression of potential difference.

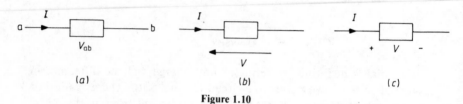

Figure 1.10

In some instances it is appropriate to relate voltage to current direction without reference to a diagram. The term voltage drop is used for this purpose and defined as the potential of the terminal where the current enters the device with respect to the terminal where it leaves. Interpreting figure 1.10(*a*) using this terminology means that V_{ab} would be described as a voltage drop. Voltage rise can similarly be defined to express V_{ba}, the potential of the terminal where the current leaves the device with respect to the terminal where it enters. In these specific terms the voltage rise for a source and the voltage drop for a sink are both positive for positive current; see figure 1.9.

Where it is necessary to specify the sense or polarity and the magnitude of a potential difference precisely, one of the methods shown in figure 1.10 is used.

1.3.3 Electromotive force

The process model of electromotive force (EMF) is that it is the electrical force which causes charge to move around a system in which energy transfer occurs. EMFs are generated within system components by various means, such as the conversion of chemical energy in a battery or changing magnetic fields in inductive devices.

When positive charge moves in the same direction as an EMF it absorbs energy from the agency producing the EMF, which is therefore a source. Conversely, if positive charge moves in the opposite direction to an EMF energy is extracted from the charge and the agency producing the EMF is a sink. Hence within a device the terminal towards which the EMF acts is the high potential one. To cause current to flow through a sink an external EMF is required.

The EMF produced within a pure circuit element is equal to its terminal voltage. For other devices the relationship between an internally generated EMF and the terminal voltage will depend on the device and its operating conditions.

The unit of EMF is the volt (V).

1.3.4 Power
Power, P, is the rate at which a charge, Q, gains or loses energy, W, in moving between two points with a voltage between them. For a steady voltage, V, the energy is given by

$$W = VQ \quad \text{and} \quad P = \mathrm{d}W/\mathrm{d}t$$

hence $P = V\mathrm{d}Q/\mathrm{d}t = VI$ watt.

If both v and i are changing then the instantaneous power, p, is given by

$$p = vi \text{ watt.}$$

The unit of power is the $\mathrm{J\,s^{-1}}$ or watt (W).

1.3.5 Convention
Strict adherence to current and voltage conventions enables the sign of P to be used to indicate whether a device is a source or a sink. If the product of current and voltage drop for a circuit element is positive, then power is being absorbed; if the product is negative the element is a source and supplying power. Alternatively, the product of the current and voltage rise for a source is positive.

1.3.6 Potential reference
The concept of potential difference suggests the difference between two values with a common reference. Thus

$$V_{\mathrm{ab}} = V_{\mathrm{ar}} - V_{\mathrm{br}}$$

where 'r' is the common reference point to which the potentials of points a and b are referred. The potential of the reference point itself is not relevant to the value of the potential difference. A general reference system takes the potential of the *earth* and any point connected to it as *zero*. All other points can then be given a positive or negative value of potential relative to the common zero reference.

1.3.7 Potential uniqueness, Kirchhoff's voltage law (KVL)

It follows from the property of electric potential that any point in a circuit has a unique voltage relative to a particular reference. The total change in potential around a closed path must be zero otherwise the point which marked the beginning and end of the path would have more than one voltage relative to the circuit reference. In energy terms this would indicate a net gain or loss of energy for any charge completing the closed path. Around a closed path any rise in potential must therefore be balanced by a corresponding fall.

This aspect of electric potential is described in various ways, generally referred to as Kirchhoff's voltage law (KVL).

To allow a more formal demonstration of KVL, the following terms are defined.

(i)   A *node* is a circuit junction to which two or more elements are connected.
(ii) A *branch* of a circuit is a single element connecting the nodes at each end of it.
(iii) A *loop* is a closed path consisting of connected branches in which no node occurs more than once.

Using this terminology the circuit shown in figure 1.11 has one loop consisting of four branches connecting the four nodes (a, b, c, d).

The total change in potential around the loop can be determined by adding the voltages across the branches in sequence. The sign of the voltages takes into account whether the potential rises or falls between successive nodes.

The circuit and graphs in figure 1.11 illustrate KVL. The bar graph shows the voltage of each node relative to the circuit reference voltage. The line graph joining the bars shows the rise and fall of potential around the loop. The equation in figure 1.11 demonstrates one form of the sequential summation of the branch voltages.

An alternative expression is obtained by reversing the relativity for each branch voltage, i.e.

$$V_{ab} + V_{bc} + V_{cd} + V_{da} = 0$$

the voltages in this equation being the negative of those in given in figure 1.11.

Program filename 'KVL' develops figure 1.11 to illustrate the properties of electric potential as follows:

(i)   the resultant change in potential around a closed path is zero;
(ii) changing the reference node moves the potential distribution graph up and down the potential axis, changing the starting point moves the graph along the position axis, both without altering its shape; and
(iii) adding voltages sequentially for a closed path produces a total of zero.

The program additionally provides practice in specifying voltages using the convention given in figure 1.10.

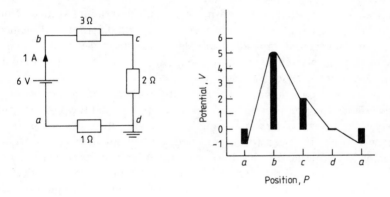

$$V_{ba} + V_{cb} + V_{dc} + V_{ad} = 0$$

$$6 - 3 - 2 - 1 = 0$$

Figure 1.11

⟩ 1.4   Passive components

1.4.1   General properties

Five basic circuit elements are used to model electronic systems; these are resistance, inductance and capacitance and voltage and current sources. In energy terms the five elements represent a sink, two storage elements and two models of an energy source, respectively.

Since resistance, inductance and capacitance represent components in which there is no net gain of energy but only loss or storage, they are collectively referred to as passive elements. Sources are used to model devices which are capable of supplying energy to a system and are therefore referred to as active elements and these are considered in §1.8.

The process models of resistance, inductance and capacitance are based on conduction, magnetic and electric fields, respectively. Field effects necessarily occupy regions of varying size but the circuit models concentrate them at a point and are consequently called lumped elements. This simplification may make lumped element models inadequate for systems in which the wavelength of any voltage or current is comparable with the size of the circuit, i.e. very large circuits operating at low frequencies or generally for circuits operating at very high frequencies. Lumped element models are assumed to be appropriate for the circuits covered in this book.

Each of the five basic elements is considered to have a pair of terminals between which a voltage drop exists and current flows. The model is defined in terms of the relationship between the terminal voltage drop and current. The elements are represented in circuit diagrams by specified symbols and the diagram shows the interconnection of circuit elements by joining the element symbols to the diagram nodes by lines representing zero resistance (see §1.9.2).

Actual physical components used to construct electronic circuits may be designed to optimize one of the above properties and the pure element may therefore be a reasonable model of them. Other devices such as integrated circuits need to be modelled in terms of many and various elements. Component technology is an important part of the electronics industry and the specification of appropriate components with respect to properties and cost are significant aspects of system design and production. Typical examples of electronic components are shown in figures 1.12–1.14 (reproduced by permission of RS Components Ltd).

Circuit analysis is based on the voltage–current relationships of the pure elements. The physical operation of each system component is a complex process and the process models given below, of how the voltage–current relationships arise, are only intended as a background for understanding the subsequent theory.

1.4.2 Resistance

The process model of resistance is based on the conduction field which exists in a conducting material when current flows in relation to an EMF applied from an external source of energy. If a continuous supply of energy is required to maintain the current flow then the material is said to have resistance.

Figure 1.12 Typical examples of resistance devices.

Figure 1.13 Typical examples of inductance devices.

Figure 1.14   Typical examples of capacitance devices.

Materials conducting current under the same physical conditions are referred to as conductors if the current is high and as insulators if it is low. Interpreting these relative terms for perfect conductors and insulators would mean infinite and zero current, corresponding to zero and infinite resistance, respectively.

The energy associated with current flow is converted by the conducting material to heat and hence is lost to the electrical system. The circuit element model of lumped resistance models the energy lost in conduction in terms of voltage. The model of lumped resistance is defined by the terminal ratio voltage drop/current.

Conductance
Conductance models the same process as resistance. The model of lumped conductance is defined by the terminal ratio current/voltage drop.

1.4.3   Inductance
The process model of inductance is based on the magnetic field associated with current flow in a conductor. Energy required to establish the

magnetic field is stored and can be recovered by the electrical system as the field decreases to zero.

The magnetic field flux links the current and inductance is defined for the process model by the ratio magnetic flux linkage/current.

The circuit element model of lumped inductance models the induced EMF as the field changes in terms of voltage. The model of lumped inductance is defined by the terminal ratio voltage drop/rate of change of current.

1.4.4 Capacitance

The process model of capacitance is based on the electric field associated with the charge transferred between two conductors which are insulated from each other. The energy required to establish the electric field is stored and can be recovered by the electrical system as the field decreases to zero.

The energy required to separate the transferred charge is modelled in terms of the voltage between the two conductors and capacitance is defined for the process model by the ratio charge transferred/voltage. The charge transferred is equal to the electric flux in the insulator (dielectric) between the two conductors.

The circuit element model of lumped capacitance models the change of charge as the field changes in terms of current. The model of lumped capacitance is defined by the terminal ratio current/rate of change of voltage drop.

1.4.5 Element parameters

The lumped element models have been previously defined in §§1.4.2–1.4.4. Table 1.1 shows the symbol used to represent, the unit for quantifying, and the defining and voltage–current relationship for each of the passive lumped elements.

The voltage–current relationships for inductance and capacitance are expressed in both differential and average form, where the average values of voltage (V_{av}) and current (I_{av}) are defined by the equations

$$V_{av} = \left(\int_{t = t_1}^{t = t_2} V_{ab} \, dt \right) (t_2 - t_1)^{-1}$$

$$I_{av} = \left(\int_{t = t_1}^{t = t_2} i \, dt \right) (t_2 - t_1)^{-1}$$

t_1 and t_2 being the time limits over which the average value is taken.

The average forms of the voltage–current relationships correspond to the process models of inductance and capacitance since

$$V_{av}t = \text{magnetic flux linkage}$$

$$I_{av}t = \text{charge or electric flux.}$$

The models of resistance, inductance and capacitance defined above are not dependent on the sign or magnitude of the defining quantities, i.e. the parameters, R, G, L and C are constants determined by the defining ratio for any corresponding values of voltage drop and current. Elements so defined are bilateral and linear.

Table 1.1

Element	Symbol	Unit	Defining and voltage–current relationship
Resistance	i R a→☐—b	ohm (Ω)	$R = V_{ab}/i$ $V_{ab} = Ri$
Conductance	i G a→☐—b	siemen (S)	$G = i/V_{ab}$ $i = GV_{ab}$
Inductance	i L a→⌒⌒⌒—b	henry (H)	$L = V_{ab}/(di/dt)$ $V_{ab} = Ldi/dt$ $i = V_{av}t/L$
Capacitance	i C a→⊣⊢—b	farad (F)	$C = i/(dV_{ab}/dt)$ $i = CdV_{ab}/dt$ $V_{ab} = I_{av}t/C$

The diode is an example of a resistive device which is neither bilateral nor linear, its resistance depending on the sign and magnitude of the voltage drop across its terminals. It is used to introduce non-linear analysis in Chapter 6. Elsewhere in this book circuit elements are bilateral and linear. A further statement is made on linearity in relation to sources in §2.2. Examples of other non-linear circuit components are inductors with ferromagnetic cores; these have an inductance which varies with current magnitude.

1.4.6 Modelling real components
Although system components such as resistors, inductors and capacitors

are designed to be working realizations of resistance, inductance and capacitance they will not consist entirely of the pure property. In some instances they may therefore need to be more precisely modelled by a combination of more than one element. Figure 1.15 shows typical representations of devices which need more than one pure element to model them accurately.

An inductor consisting of a coil of wire with significant resistance can be modelled by a series combination of inductance and resistance. A capacitor with a non-perfect dielectric can be modelled by a parallel combination of capacitance and resistance.

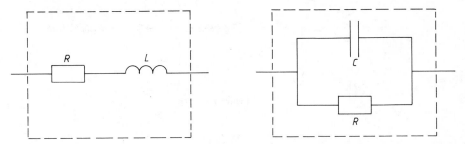

Figure 1.15

1.4.7 Mutual inductance

Mutual inductance is based on the process model of two inductances with a proportion of their magnetic fields in common, as shown in figures 1.16 and 1.17. Such inductances are described as being magnetically coupled and additionally possess mutual inductance, M. Since part of the field is common to both elements either or both can supply or recover energy from it and both inductances have additional voltages associated with EMFs induced by the changing field. The model of lumped mutual inductance is defined by the terminal ratio voltage drop for one inductance/rate of change of current in the other. Note that this definition only applies when the inductance for which the voltage drop is specified has no voltage drop due to its own inductance, for example, if in the equation for V_{ab} below, di_1/dt was equal to zero.

The unit of mutual inductance is the henry (H).

The voltage equations for the two inductances shown in figure 1.16 are

$$V_{ab} = L_1(di_1/dt) + M(di_2/dt)$$

$$V_{cd} = L_2(di_2/dt) + M(di_1/dt).$$

The sign which precedes the voltage term due to mutual inductance depends on the relative directions of the fields due to each current.

Where it is necessary to specify the sign of the voltages due to mutual inductance the corresponding terminals of the two elements can be labelled with a dot to relate the current and field directions. The convention adopted is that currents of the same sign flowing in the same direction relative to the dot produce fields in the same direction. Thus if i_1 and i_2 in figure 1.16 are both positive, then the resultant magnetic field is greater than the field due to either current alone. For inductances labelled as in figure 1.16 the voltage equations are as given above.

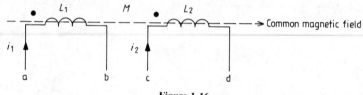

Figure 1.16

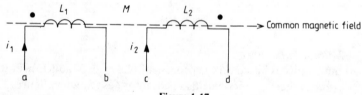

Figure 1.17

Figure 1.17 with the related voltage equations shows the effect of reversing the direction of one of the currents relative to the dot terminal,

$$V_{ab} = L_1(\mathrm{d}i_1/\mathrm{d}t) - M(\mathrm{d}i_2/\mathrm{d}t)$$

$$V_{cd} = L_2(\mathrm{d}i_2/\mathrm{d}t) - M(\mathrm{d}i_1/\mathrm{d}t).$$

Equivalent 'T' circuit

The coupled inductances in figure 1.18 can be modelled by the equivalent 'T' circuit shown.

The terminal voltages for the 'T' circuit are given by the equations

$$V_{ab} = (L_1 - M)(\mathrm{d}i_1/\mathrm{d}t) + M[\mathrm{d}(i_1 + i_2)/\mathrm{d}t]$$

$$V_{cd} = (L_2 - M)(\mathrm{d}i_2/\mathrm{d}t) + M[\mathrm{d}(i_1 + i_2)/\mathrm{d}t].$$

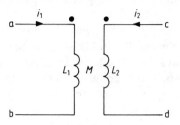

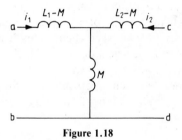

Figure 1.18

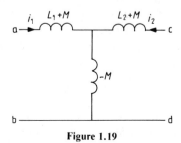

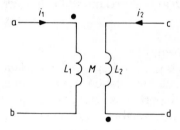

Figure 1.19

These equations simplify to those given for the coupled inductances in figure 1.16.

Figure 1.19 shows the equivalent 'T' circuit for the alternative condition corresponding to the inductances in figure 1.17.

Although the equivalent 'T' circuits produce the same voltage equations as the corresponding coupled inductances they represent, in another respect the systems that the circuits model are electrically significantly different. In the 'T' circuits the terminals b and d are connected together, whereas the coupled inductances are electrically isolated and transfer energy through the common magnetic field.

The property in actual systems corresponding to the mutual inductance model is frequently used for isolating circuits electrically while maintaining an energy link between them. This is the physical basis of one of the most common electrical devices, namely the transformer, and is generally a widely employed physical property. It is also the cause of unwanted interference between electronic systems which are unintentionally linked in this manner and frequently appropriate shielding is required to prevent it.

⟩ 1.5   Voltage–current ($v-i$) relationships for passive components

1.5.1   Common waveforms

Figure 1.20 shows how the voltage–current relationships of the three passive elements operate in relation to some common waveforms and also illustrates some important general points. For resistance, the current and voltage waveforms are always the same shape with the same time phase. With inductance and capacitance the differential relationship produces a change in waveform for the square and triangular waveforms and a phase difference for the sine waves.

The term voltage–current when applied to the element relationships means voltage in terms of current, or current in terms of voltage, whichever is more appropriate in the context. The term is frequently abbreviated to $v-i$ in the text.

1.5.2   Relationship between current and voltage maxima

For the waveforms shown in figure 1.20 the ratio between the instantaneous values of voltage and current is constant and equal to R for the resistance but varies between plus and minus infinity for the inductance and capacitance.

Square and triangular waveforms
The triangular waveform has a rate of change which is plus or minus a constant value. Triangular waveforms of current and voltage are therefore associated with square waves of voltage and current for inductance and capacitance, respectively.

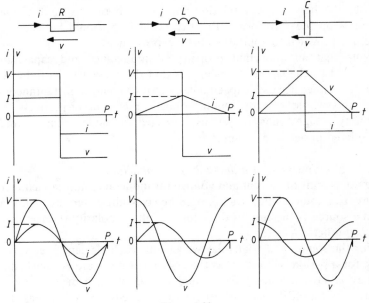

Figure 1.20

The maximum values are related as follows:

$$V = L[I/(P/2)]$$

for the inductance and

$$I = C[V/(P/2)]$$

for the capacitance, where P is the periodic time of the waveforms.

Sine waves
The rate of change of a sine wave is a cosine wave. Sine waves of current

and voltage are therefore associated with cosine waves of voltage and current for inductance and capacitance, respectively.

The maximum values are related as follows:

$$V = L(2I\pi/P)$$

for the inductance and

$$I = C(2V\pi/P)$$

for the capacitance; see Chapter 3 for the derivation of this relationship.

Program filename 'VIC' produces figure 1.20 sequentially for the user to observe how the $v-i$ relationship for each of the three elements operates on the waveforms. Practice problem 5 should be used as an example in which to apply the above expressions.

Note that an important property of inductance and capacitance, illustrated by the $v-i$ relationships, is that the current in an inductance and the voltage across a capacitance cannot be changed instantaneously, since this would require infinite voltage or current respectively. This property is significant in systems subjected to sudden changes and is considered in detail in Chapter 5.

1.5.3 Energy stored in inductance and capacitance

The waveforms of current and voltage for inductance and capacitance in figure 1.20 indicate that each device behaves alternately as a sink and then a source as the current direction or voltage polarity reverse relative to each other.

For the square and triangular waveforms each element operates as a sink for a period of time $\frac{1}{2}P$ as the associated fields are established and energy is stored. During the second half cycle each device acts as a source; as the fields decrease to zero the stored energy is released. By reference to these waveforms the relationships for stored energy can be derived as follows.

A parameter which increases linearly with time from zero has an average value equal to half the value reached over the time being considered. For the triangular waveforms in figure 1.20 the average values over the time $\frac{1}{2}P$ when they are increasing are therefore half the maximum values.

Inductance

The average power, P_{av}, for the time 0 to $\frac{1}{2}P$ is given by

$$P_{av} = \frac{1}{2}VI$$

and the stored energy, W, at time $\frac{1}{2}P$ by

$$W = \tfrac{1}{2}P_{av}P = \tfrac{1}{4}VIP.$$

Since the maximum voltage V is given by (see §1.5.2)

$$V = 2LI/P$$

$$W = \tfrac{1}{2}LI^2 \text{ J}.$$

Capacitance
The average power for the time 0 to $\frac{1}{2}P$ is given by

$$P_{av} = \tfrac{1}{2}VI$$

and the stored energy by

$$W = \tfrac{1}{4}VIP.$$

Since the maximum current I is given by

$$I = 2CV/P$$

$$W = \tfrac{1}{2}CV^2 \text{ J}.$$

Although these relationships are derived from specific waveforms the stored energy at any instant is only dependent on the state of the associated field and not how that state was achieved.

For inductance the strength of the magnetic field is determined by the current value and for capacitance the strength of the electric field by the voltage. Hence the above expressions are generally applicable for determining the stored energy for any value of current or voltage.

The storage property of inductance and capacitance is used in various applications over a range of power levels. It is also frequently necessary to consider stored energy when changing the state of a system. For example, electronic devices may be damaged when the current in inductive systems is rapidly reduced to zero if no means has been provided for dissipating the stored energy released by the field.

1.5.4 Power in passive components

Inductance and capacitance
As described in the previous subsection, under alternating voltage and current conditions inductance and capacitance behave alternately as sinks and then as sources of energy, as the fields associated with them are

established and then reduced to zero. It follows therefore that over a complete cycle the average power for these elements is zero.

Resistance

Resistance is always a sink of energy whether operating under unidirectional or alternating current conditions. In the latter case the current and voltage reverse together, which means the instantaneous power is always positive or zero. With reference to the parameters in figure 1.21 the power is given by

$$p = vi = i^2R = v^2/R \text{ W.}$$

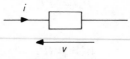

Figure 1.21

If the current and voltage are constant at I and V, respectively, then the power p is also constant and

$$p = VI = I^2R = V^2/R \text{ W.}$$

If the current and voltage are varying the average power

$$P_{av} = \text{mean of } (i^2R) = \text{mean of } (v^2/R) \text{ W.}$$

Since R is constant

$$P_{av} = [\text{mean of } (i^2)]\,R = [\text{mean of } (v^2)]/R$$

and the average power can be expressed in terms of the square root of the mean of i^2 or v^2, the root-mean-square (RMS) values. Hence

$$P_{av} = (I_{rms})^2R = (V_{rms})^2/R \text{ W.}$$

The RMS value of a sine wave is covered in §3.2.3.

It is apparent from the power relationships that the RMS values are equivalent to the steady values with respect to power dissipating properties.

⟩ 1.6   Series and parallel equivalents

Series

Components connected in series have one common exclusive connection. With no other conducting paths at the common connection, components connected in series must carry the same current. The equivalent single component for series combination is found by substituting the resultant voltage in the appropriate $v-i$ relationship, as shown in figure 1.22.

The equivalent inductance of two inductances connected in series is determined in figure 1.22 on the assumption that no mutual inductance exists between the two inductances. The equivalent inductance for the series connection of two inductances with mutual inductance is shown in figure 1.23.

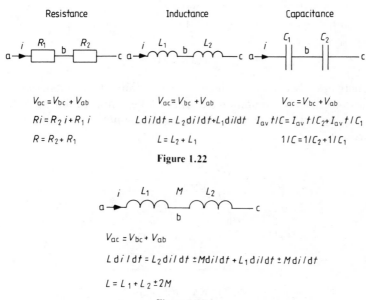

Resistance   Inductance   Capacitance

$V_{ac} = V_{bc} + V_{ab}$ $V_{ac} = V_{bc} + V_{ab}$ $V_{ac} = V_{bc} + V_{ab}$

$Ri = R_2 i + R_1 i$ $L di/dt = L_2 di/dt + L_1 di/dt$ $I_{av} t/C = I_{av} t/C_2 + I_{av} t/C_1$

$R = R_2 + R_1$ $L = L_2 + L_1$ $1/C = 1/C_2 + 1/C_1$

Figure 1.22

$V_{ac} = V_{bc} + V_{ab}$

$L di/dt = L_2 di/dt \pm M di/dt + L_1 di/dt \pm M di/dt$

$L = L_1 + L_2 \pm 2M$

Figure 1.23

Parallel

Components connected in parallel have two common connections and hence the same voltage across them. The equivalent single component for parallel combination is found by substituting the resultant current in the appropriate $v-i$ relationship, as shown in figure 1.24.

The parallel combination of inductances shown in figure 1.24 assumes no mutual inductance. The parallel combination of inductances with mutual inductance is more complex than the series case and is given in Appendix A1.6.

Program filename 'CC' uses the relationships derived in figures 1.22 and 1.24 to compute the series and parallel combination of the elements

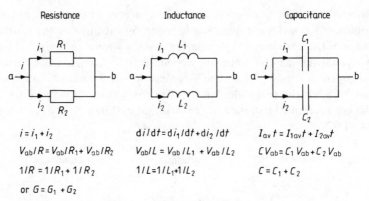

Resistance	Inductance	Capacitance

$i = i_1 + i_2$

$V_{ab}/R = V_{ab}/R_1 + V_{ab}/R_2$

$1/R = 1/R_1 + 1/R_2$

or $G = G_1 + G_2$

$di/dt = di_1/dt + di_2/dt$

$V_{ab}/L = V_{ab}/L_1 + V_{ab}/L_2$

$1/L = 1/L_1 + 1/L_2$

$I_{av}\,t = I_{1av}t + I_{2av}t$

$CV_{ab} = C_1 V_{ab} + C_2 V_{ab}$

$C = C_1 + C_2$

Figure 1.24

in a network where all the elements are of a similar type and in the case of inductance, *without mutual inductance*. An example to illustrate the use of the program together with the resulting computer printout is given in figure 1.25.

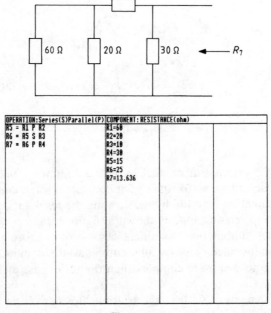

OPERATION:Series(S)Parallel(P)	COMPONENT: RESISTANCE(ohm)			
R5 = R1 P R2	R1=60			
R6 = R5 S R3	R2=20			
R7 = R6 P R4	R3=10			
	R4=30			
	R5=15			
	R6=25			
	R7=13.636			

Figure 1.25

Practice problems 4 and 6 and other similar examples should be solved by using the program and used to gain experience in describing circuit structure in terms of series and parallel connection.

⟩ 1.7   Voltage and current division

1.7.1   Voltage division

Since resistances connected in series carry the same current the total voltage across a series combination divides in direct proportion to the resistances to produce the voltage across each element. Referring to the resistances in figure 1.22,

$$V_{bc}/V_{ac} = R_2 i/[(R_2 + R_1)i] = R_2/(R_2 + R_1)$$

$$V_{ab}/V_{ac} = R_1 i/[(R_2 + R_1)i] = R_1/(R_2 + R_1).$$

The sliding contact resistor shown in figure 1.26 uses this principle to produce a continuously variable voltage between the sliding contact and one end of it. This device is used widely as a volume control in audio equipment and many other similar applications.

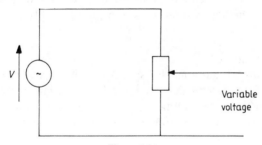

Variable
voltage

Figure 1.26

1.7.2   Current division

Since resistances connected in parallel have the same voltage across them, the total current divides in proportion to the conductances to produce the current in each element.

For the resistances shown in figure 1.24, $G_1 = 1/R_1$ and $G_2 = 1/R_2$ hence

$$i_1/i = G_1 V_{ab}/[(G_1 + G_2)V_{ab}] = G_1/(G_1 + G_2) = R_2/(R_1 + R_2)$$

$$i_2/i = G_2 V_{ab}/[(G_1 + G_2)V_{ab}] = G_2/(G_1 + G_2) = R_1/(R_1 + R_2).$$

Voltage and current division are frequently used as techniques for monitoring large voltages and currents by sampling a smaller known proportion of the larger quantity. Practice problem 6 is an example of this application.

> ### 1.8 Sources

1.8.1 General principles

Energy sources in electric circuits may represent input to the system or may be part of the model of an electronic device within it. In the latter case the device may be controlling the flow of energy from elsewhere in the system and the source therefore modelled as a *controlled* and not an *independent* one.

Primary supplies of energy to an electronic system require the conversion of energy of another type, such as chemical in a battery or mechanical/thermal in a power station. Low power and/or portable electronic equipment may have its own primary source. Other equipment will derive its power from the mains and the supply to any circuit in the system is therefore a secondary controlled transfer of energy. Energy input to an electric circuit is modelled as a voltage or current source regardless of whether it represents a primary supply or a controlled transfer.

1.8.2 Voltage source

A voltage source maintains the specified voltage across its terminals irrespective of the current delivered and has the characteristic shown in figure 1.27. Alternating voltage sources undergo periodic reversals of polarity and the waveform can have various envelopes such as sinusoidal, square and triangular.

For an independent source the voltage is not determined by the circuit, but for a controlled source the voltage is determined by a circuit parameter. A number of different symbols are used and some common ones are shown in figure 1.27.

1.8.3 Current source

A current source delivers the specified current regardless of the voltage across its terminals and has the characteristic shown in figure 1.27. The source current may be unidirectional (direct current, DC) or undergo periodic reversals in direction (alternating current, AC) with various waveform envelopes.

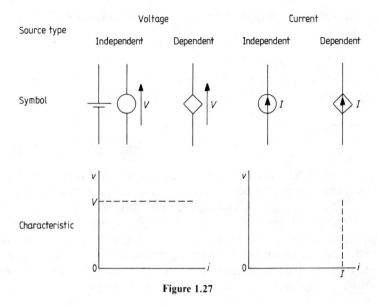

Figure 1.27

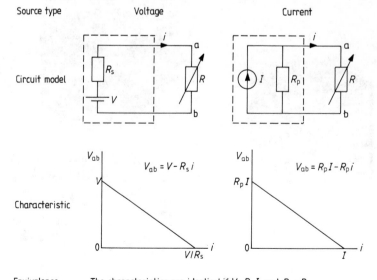

Figure 1.28

Sources may be independent or dependent ones controlled by current or voltage.

1.8.4 Voltage and current generators

In a similar way to passive components the physical realization of an energy source, whether primary or secondary, may approximate to the pure element. In general the terminal characteristics of a voltage or current generator will need to be modelled as a combination of elements. Two simple but common generator models are shown and compared in figure 1.28.

The terminal characteristics of generators which can be modelled in this way can be represented by either combination and the equivalence is shown. The resistances R_s and R_p are referred to as internal resistances since they are within the terminals of the generators. The generator approaches the pure voltage or current source as R_s and R_p approach zero and infinite resistance, respectively. It should be noted that the pure elements cannot be made equivalent since R_s and R_p would both be required to approach zero and infinity at the same time.

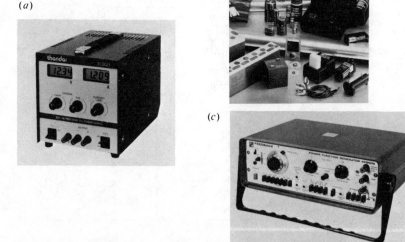

(a)

(b)

(c)

Figure 1.29 (a) Bench power supply, (b) batteries and charging equipment, (c) 1 kHz power function generator. Reproduced by permission of RS Components Ltd.

Figure 1.29 shows some typical realizations of voltage generators. The internal resistance will frequently be quoted as part of the specification of such equipment.

⟩ 1.9   Limiting values

1.9.1   Pure elements, real components

The convention has been adopted of using the terms resistance, inductance and capacitance for pure elements and resistor, inductor and capacitor for the real components which correspond to optimal realizations of the models. Similarly, the terms voltage and current source are used for the pure source elements of voltage and current generators. The pure elements can thus be seen as models which represent the idealized limits of the practical components. In addition it is sometimes a necessary analytical technique to consider the limits of the pure elements themselves.

1.9.2   Zero and infinite limits of passive elements

Resistance
Zero resistance exists where conduction takes place without loss of energy. The term short circuit is frequently used for a zero resistance connection, represented in circuit diagrams by plain connecting lines. Infinite resistance means no conduction takes place or that perfect insulation exists, the corresponding term being open circuit.

Inductance
Zero inductance behaves as a short circuit since no voltage is produced however rapid the rate of change of current. Alternatively, infinite inductance behaves as an open circuit since an infinite voltage would be produced however slowly the current attempted to change, hence no current can be established.

Capacitance
Zero capacitance means no current can be established however rapid the voltage rate of change, hence it behaves as an open circuit. Infinite capacitance behaves as a short circuit since however slowly the voltage attempts to rise the current would be infinite and hence no voltage can be established.

1.9.3   Zero sources

In some analytical processes it is necessary to consider the effect of reducing a source to zero, which is not necessarily the same as removing it completely.

Zero voltage source

The characteristic of the voltage source indicates that unlimited current can flow at the specified voltage. Hence for a zero voltage source this is the same property as a short circuit and is clearly not the same as removing the source completely.

Zero current source

A current source delivers its current regardless of the voltage across its terminals. Thus a zero current source has the characteristic of an open circuit and is equivalent to removing it completely.

〉1.10   Simple application of Kirchhoff's laws

Kirchhoff's two laws together provide the basis for much circuit analysis. The principle of their application to the solution of circuit branch currents is demonstrated on the simple circuit shown in figure 1.30. More sophisticated methods and analysis techniques are described in later chapters.

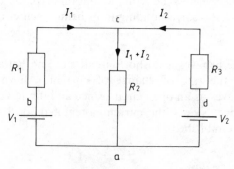

Figure 1.30

The particular problem addressed in this section is to determine the branch current in a given circuit with known voltage sources and resistances. Although the circuit has five elements there are only three different values of current, hence three equations are required for their solution. One equation is used in applying the current law at node c to label the circuit currents. This is a useful general technique because it reduces the remaining number of equations to be manipulated.

The remaining two equations are produced by applying the voltage law to the two loops abca and acda as shown below. For loop abca

$$V_{ba} + V_{cb} + V_{ac} = 0$$

and loop acda

$$V_{ca} + V_{dc} + V_{ad} = 0$$

hence

$$V_1 - R_1 I_1 - R_2(I_1 + I_2) = 0$$

and

$$R_2(I_1 + I_2) + R_3 I_2 - V_2 = 0$$

These equations can be rearranged to give

$$V_1 = (R_1 + R_2)I_1 + R_2 I_2$$

$$V_2 = R_2 I_1 + (R_2 + R_3)I_2.$$

With V_1, V_2, R_1, R_2 and R_3 being known these equations can be solved to give the value of I_1 and I_2. Appendix A1.10 illustrates a rule for solving equations of this form.

SOLUTION OF TWO SIMULTANEOUS EQUATIONS

The two equations should be in the form:

U1 = R1.I1 + R2.I2

U2 = R3.I1 + R4.I2

I1 = 0.508

I2 = -9.52E-2

Figure 1.31

Program filename 'KLA' is a program for solving two simultaneous equations. A numerical example of the circuit in figure 1.30 where V_1, V_2, R_1, R_2 and R_3 are 4 V, 2 V, 3 Ω, 6 Ω and 5 Ω respectively produces the printout shown in figure 1.31.

Note that the coefficients of the unknown currents in the computer program do not necessarily correspond to actual circuit resistances. Practice problem 3 provides further examples for using this method and program.

⟩ 1.11   Practice problems

1   Determine the electron velocity in copper wire carrying one ampere if the diameter is (i) 1 mm and (ii) 0.1 mm. It can be assumed that the number of free electrons for conduction in copper is 11×10^{22} cm^{-3}.

2   A VDU tube has a beam current of 1 mA. The electrons travel at 2×10^7 m s^{-1} and the beam is circular with a diameter of 1 mm. Determine the density of electrons in the beam.

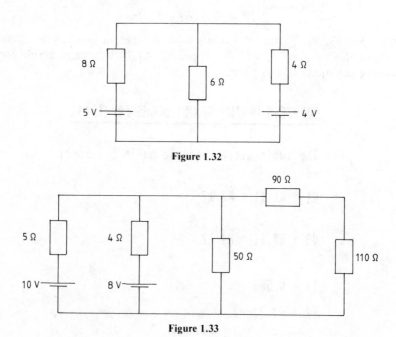

Figure 1.32

Figure 1.33

3 Use Kirchhoff's laws to determine the current and power in each element of the circuit shown in figure 1.32 and the current in each resistance in the circuit shown in figure 1.33.

4 For the circuits shown in figures 1.32 and 1.33 determine the total resistance in series with each source if the other source is reduced to zero.

5 For the waveforms shown in figure 1.20 determine the maximum values of the current waveforms if the maximum voltage is 1 V and the frequency is 1000 Hz in each case. The values for the resistance, inductance and capacitance are 1000 Ω, 1×10^{-3} H and 1×10^{-6} F, respectively.

6 A sensitive analogue ammeter can be extended to indicate a range of voltages and currents by having switched series and parallel resistances as shown in figure 1.34. The series resistances are called multipliers and the parallel ones shunts. Such an ammeter has a resistance of 5 Ω and produces a full scale deflection when passing a current of 5×10^{-5} A. Using the relationships given in §1.7, determine:

(i)   the multipliers required to give full scale ranges of 1, 10 and 100 V;
(ii) the shunts required for 1 mA, 1 A and 10 A current ranges;
(iii) the overall resistance of the instrument on all the ranges.

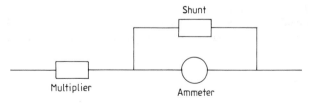
Figure 1.34

7 The circuits shown in figure 1.35 have a particular relationship which is the subject of §2.2. Determine the currents I_1, I_2 and I_3.

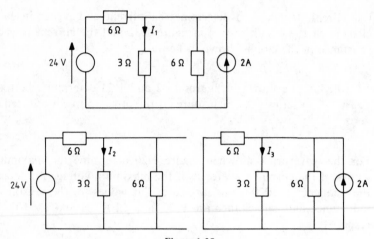

Figure 1.35

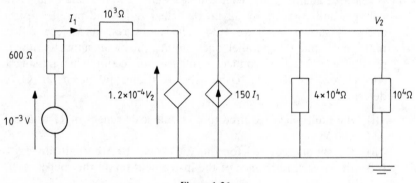

Figure 1.36

8   For the circuit shown in figure 1.36 determine the voltage, V_2, across the 1×10^{-4} Ω resistance and the required power rating for this component.

> Chapter 2

> Analytical Techniques and Processes

> 2.1 Introduction

In §1.5 it was shown that resistance only produces scaling of voltage and current waveforms, whereas inductance and capacitance can change the shape or the phase of one waveform in relation to the other. It is clear therefore that the more sophisticated aspects of signal processing or other applications of circuit theory will involve inductance and capacitance in alternating or changing voltage and current systems.

There are a number of general techniques and processes which can be used to analyse circuits and, within the framework of the above constraint, the approach adopted here is to introduce some of them with resistance only circuits. Although such circuits are limited practically they are appropriate for illustrating the various processes under discussion. By avoiding the complication produced by the storage properties and differential characteristics of inductance and capacitance the process itself is brought into focus. Having demonstrated the processes on more simple circuits, the techniques required for the analysis of more complicated systems become apparent.

Since purely resistive circuits only involve voltage and current scaling, without phase changes, the same relationship governs the current and voltage at all instances in time and similar analysis techniques can be used for direct or alternating signals. The term signal is used here and throughout the text as a generic one to mean a circuit current or voltage, without reference to magnitude or application. In some instances this may coincide with the more specific meaning of signal, which is a current or voltage conveying information.

The circuits used to demonstrate particular aspects of the processes are chosen for illustration, more common practical structures being left for practice problems.

⟩ 2.2 Linearity and superposition

The principle of superposition is essentially a particular expression of the property of linearity since a linear system embraces the concept of superposition. In broad terms a system is linear when effect is directly proportional to cause. For an electric circuit this means, for example, that if a voltage sourse is doubled then the current response it produces will also be doubled. The particular expression of this effect, which is called the principle of superposition, applies to circuits with more than one source. In such circuits, the total response in any part is the sum of the responses which would be produced by each source acting alone, all other sources having been reduced to zero.

Reference to §1.9.3 indicates that reducing a voltage source to zero means replacing it with a short circuit and reducing a current source to zero is equivalent to replacing it with an open circuit. The principle of superposition is illustrated using the circuits shown in figure 2.1.

Program filename 'SUPRPOS' reproduces figure 2.1 sequentially to further demonstrate the principle and then uses practice problem 1.7 as a

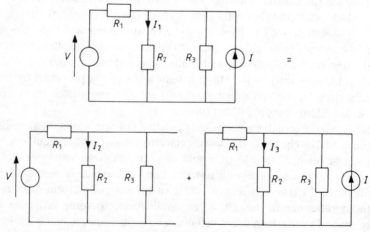

Figure 2.1 $I_1 = I_2 + I_3$.

numerical example. The example values are chosen to allow an easy check of the result to confirm the analysis.

The principle of superposition has a wider and more general use than indicated above. In addition to the type of application shown it can be used in the reverse sense. Examples of this use are resolving a sinusoidal signal into quadrature components or a non-sinusoid into harmonic ones. The total response to such signals is then produced by summing the component responses. These applications will be covered in later chapters.

It is important to note that, because power depends on the square of current or voltage, the principle of superposition cannot be applied to power calculation. The current and voltage used to determine the power dissipated in various parts of a circuit must be the total values for the appropriate elements.

⟩ 2.3   Mesh analysis

Mesh analysis is a systematic application of Kirchhoff's laws to determine the currents in a circuit with known voltage sources and resistances.

The term mesh applied to a circuit means a space surrounded by a continuous conducting path. The continuous path consists of circuit elements and a complete circuit is a network of such meshes. More formally, a mesh is a loop as defined in §1.3.7, which does not contain any other loops within it.

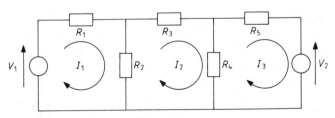

Figure 2.2

The network shown in figure 2.2 has three meshes each of which is ascribed a circulating current. For a completely systematic approach, which lends itself to computer solution, the mesh currents are all given the same direction, in this case clockwise. The mesh circulating current is the actual current in any element of the mesh which is not common with other meshes. In elements which are common to two meshes, the current is the difference between the two mesh currents involved. Since there are three meshes in the circuit, three simultaneous equations are required for the solution of the corresponding currents.

In general, the number of equations required is equal to the number of meshes. These are obtained by applying Kirchhoff's voltage law to each mesh in turn. The sum of the voltages added in sequence in a clockwise direction around each mesh is equated to zero as shown.

Mesh 1 $\qquad V_1 - R_1 I_1 - R_2(I_1 - I_2) = 0$

Mesh 2 $\qquad - R_2(I_2 - I_1) - R_3 I_2 - R_4(I_2 - I_3) = 0$

Mesh 3 $\qquad - R_4(I_3 - I_2) - R_5 I_3 - V_2 = 0$

These equations can be rearranged to equate the known resultant voltage sources acting *clockwise* in each mesh to the sum of the voltages produced by the unknown mesh currents flowing in the circuit resistances:

$$V_1 = (R_1 + R_2)I_1 - (R_2)I_2$$

$$0 = - (R_2)I_1 + (R_2 + R_3 + R_4)I_2 - (R_4)I_3$$

$$- V_2 = - (R_4)I_2 + (R_4 + R_5)I_3$$

Note that the total voltage source in any mesh acting in the direction of the mesh current is the sum of the voltage rises of all the sources in the mesh in relation to the direction of the particular mesh current (see §1.3.2).

2.3.1 Matrix expression
The above equations can further be expressed in matrix form, $[V] = [R][I]$, shown below (the process of matrix multiplication is given in Appendix A2.3):

$$\begin{bmatrix} V_1 \\ 0 \\ - V_2 \end{bmatrix} = \begin{bmatrix} R_1 + R_2 & - R_2 & 0 \\ - R_2 & R_2 + R_3 + R_4 & - R_4 \\ 0 & - R_4 & R_4 + R_5 \end{bmatrix} \begin{bmatrix} I_1 \\ I_2 \\ I_3 \end{bmatrix} .$$

Inspection of the matrix equation shows four types of entry as follows.

(i) For the voltage matrix each element is the total voltage source in each mesh acting in the direction of the mesh current, V_1 for mesh 1, 0 for mesh 2 and $- V_2$ for mesh 3.

(ii)   For the resistance matrix, on the diagonal, the total resistance in each mesh is $(R_1 + R_2)$ for mesh 1, $(R_2 + R_3 + R_4)$ for mesh 2 and $(R_4 + R_5)$ for mesh 3.

(iii) The remaining elements in the resistance matrix are the total resistances common to the two appropriate meshes prefixed by a negative sign. The negative sign arises from the fact that in elements common to two meshes the mesh currents flow in opposite directions. Hence the voltages in the common elements due to the two mesh currents are in opposite senses. The voltage due to the current of the mesh being considered has a positive sign, while voltages due to other mesh currents in common elements have negative signs.

   The elements are $(-R_2)$ for the resistance common to meshes 1 and 2, $(-R_4)$ for the resistance common to meshes 2 and 3 and (0) since there is no common resistance between meshes 1 and 3.

(iv)   For the current matrix the elements are the clockwise mesh currents I_1, I_2 and I_3.

It is particularly important when using an automatic problem-solving technique such as mesh analysis that the principles of the process are understood before it is used automatically.

Program filename 'MESH' illustrates the process of producing the matrix equation sequentially and then gives a numerical example. The sequential process shows the contribution that each mesh current makes to the voltage equation for the mesh being considered. The numerical example provides an opportunity for writing down the matrix equation before it is checked against the program.

Once the process of producing the matrix equation has been understood, programs under filename 'MA' can be used to solve the numerical example and for further practice and other problems. The numerical example and the resulting computer printout of the solution are shown in figure 2.3. The problem-solving programs allow for input in different forms as the menu describes and use two different numerical methods. They are arranged in order of progressive generality to match increasing user familiarity with this analysis technique. Each program provides an opportunity for changing the data before a recalculation is performed. This facility can be used to calculate the effect of varying specific component values.

A brief description of the numerical methods used is given in Appendix A2.3.

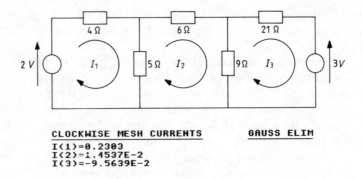

CLOCKWISE MESH CURRENTS   GAUSS ELIM
I(1)=0.2303
I(2)=1.4537E-2
I(3)=-9.5639E-2

CLOCKWISE MESH CURRENTS It.no.=11
I(1)=0.2303
I(2)=1.4537E-2
I(3)=-9.5639E-2

Figure 2.3

2.3.2 Dependent sources

As stated in Chapter 1, dependent sources are frequently required for the simulation and analysis of electronic circuits. The principles of mesh analysis of circuits containing dependent sources are the same as those described in §2.3 but the resulting matrix equation becomes modified.

· The circuit shown in figure 2.4 has a similar structure to that in figure 2.2 but the independent voltage source V_1 has been replaced by a dependent one, RI_3, whose voltage is directly proportional to the mesh current I_3.

Using the standard technique on the circuit in figure 2.4 produces the matrix equation shown below with an unknown quantity (RI_3) in the voltage matrix.

$$\begin{bmatrix} RI_3 \\ 0 \\ -V_2 \end{bmatrix} = \begin{bmatrix} R_1 + R_2 & -R_2 & 0 \\ -R_2 & R_2 + R_3 + R_4 & -R_4 \\ 0 & -R_4 & R_4 + R_5 \end{bmatrix} \begin{bmatrix} I_1 \\ I_2 \\ I_3 \end{bmatrix}.$$

To produce a solution this unknown voltage is transferred to the other side of the equation as an extra resistance in the resistance matrix, as follows:

$$\begin{bmatrix} 0 \\ 0 \\ -V_2 \end{bmatrix} = \begin{bmatrix} R_1 + R_2 & -R_2 & -R \\ -R_2 & R_2 + R_3 + R_4 & -R_4 \\ 0 & -R_4 & R_4 + R_5 \end{bmatrix} \begin{bmatrix} I_1 \\ I_2 \\ I_3 \end{bmatrix}.$$

The equation can now be solved in the standard way and program filename 'MA' should be used to solve practice problem 5 and other similar problems as numerical examples of the process.

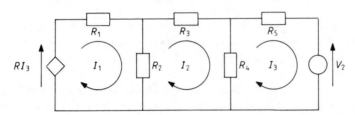

Figure 2.4

⟩ 2.4 Node analysis

Node analysis provides an alternative systematic use of Kirchhoff's laws for the solution of circuit problems. The circuit model is expressed in terms of current sources and conductances, unknown node voltages being determined in terms of these known parameters. In the first instance the process is described with independent sources and then extended to include dependent ones.

For an isolated circuit, as shown in figure 2.5, the node potentials can be determined in relation to each other or with respect to an external reference point.

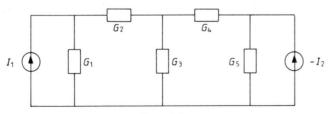

Figure 2.5

It is common practice to specify one of the circuit nodes as the potential reference and the circuit is redrawn in figure 2.6 to highlight the reference node.

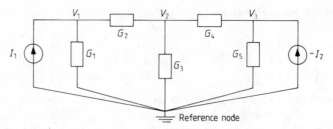

Figure 2.6

Although the circuit has four nodes, identifying one as the reference node leaves only three unknown node voltages with respect to the reference node, V_1, V_2 and V_3 respectively. In general the number of unknown node voltages is one less than the total number of nodes. The node voltages are determined from simultaneous equations produced by applying Kirchhoff's current law to each node in turn. The sums of the currents towards each node are added and equated to zero as shown.

Node 1 $I_1 - G_1 V_1 - G_2(V_1 - V_2) = 0$

Node 2 $- G_2(V_2 - V_1) - G_3 V_2 - G_4(V_2 - V_3) = 0$

Node 3 $- I_2 - G_4(V_3 - V_2) - G_5 V_3 = 0$

These equations can be rearranged to equate the known resultant current source acting towards a node to the current flowing away from it, expressed as the sum of the products of the appropriate voltages with the conductances meeting at the node. Where a conductance links the node being considered directly to the reference node, the voltage across it is the node voltage. Where the conductance links another node the voltage across it is the difference between the two node voltages:

$$I_1 = (G_1 + G_2)V_1 - (G_2)V_2$$

$$0 = -(G_2)V_1 + (G_2 + G_3 + G_4)V_2 - (G_4)V_3$$

$$- I_2 = -(G_4)V_2 + (G_4 + G_5)V_3.$$

2.4.1 Matrix expression

The above equations can be expressed in the matrix form, $[i] = [G][V]$, as shown below:

$$\begin{bmatrix} I_1 \\ 0 \\ -I_2 \end{bmatrix} = \begin{bmatrix} G_1 + G_2 & -G_2 & 0 \\ -G_2 & G_2 + G_3 + G_4 & -G_4 \\ 0 & -G_4 & G_4 + G_5 \end{bmatrix} \begin{bmatrix} V_1 \\ V_2 \\ V_3 \end{bmatrix}.$$

As with mesh analysis there are four types of matrix entry.

(i)   For the current matrix each element is the total current source acting towards each node in turn: I_1 for node 1, 0 for node 2 and $-I_2$ for node 3.

(ii) For the conductance matrix on the diagonal, the total conductance at each node is $(G_1 + G_2)$ for node 1, $(G_2 + G_3 + G_4)$ for node 2 and $(G_4 + G_5)$ for node 3.

(iii) The remaining conductance matrix elements are the total conductances linking each pair of appropriate nodes prefixed by a negative sign. The negative sign is produced because the voltage across the conductance is not the node voltage but the difference between the two node voltages involved.

      The elements are $-G_2$ for nodes 1 and 2, $-G_4$ for nodes 2 and 3 and 0 for nodes 1 and 3 since there is no direct link.

(iv) For the node voltage matrix the elements are the node voltages with respect to the reference node, V_1, V_2 and V_3.

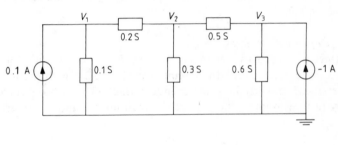

```
NODE VOLTAGES                    GAUSS ELIM
V(1)=-7.109E-2
V(2)=-0.60664
V(3)=-1.1848
```

```
NODE VOLTAGES              It.no.=17
V(1)=-7.109E-2
V(2)=-0.60664
V(3)=-1.1848
```

Figure 2.7

The program sequence for node analysis is the same as that for mesh analysis. Program filename 'NODE' illustrates the process of producing the matrix equation sequentially and then a numerical example is given. The sequential process shows the contribution that each node voltage makes to the current flow at the node being considered. This requires particular attention because it separates the contribution that the voltage of each end of a conductance, with respect to the reference, makes to the current flowing through it. This is more complicated than simply considering the current due to the voltage across the conductance but it is necessary to identify each element in the matrix equation. The numerical example provides an opportunity for writing down the matrix equation before checking it against the program.

Once the process has been understood programs under filename 'NA' can be used to solve the numerical example and further practice and other problems. The numerical example and the computer printout of the solution are shown in figure 2.7.

The numerical methods used are the same as those for the mesh analysis programs and the provision for changing data and recalculation are also similar.

2.4.2 Dependent sources
The effect produced by dependent sources is similar to that for mesh analysis. Figure 2.8 shows the circuit from figure 2.5 with the independent source I_1 replaced by a dependent one, GV_2, whose current is directly proportional to the voltage between node 2 and the reference.

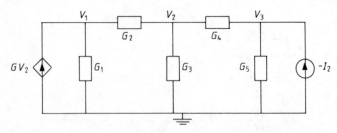

Figure 2.8

Using the standard procedure produces the following matrix:

$$\begin{bmatrix} GV_2 \\ 0 \\ -I_2 \end{bmatrix} = \begin{bmatrix} G_1 + G_2 & -G_2 & 0 \\ -G_2 & G_2 + G_3 + G_4 & -G_4 \\ 0 & -G_4 & G_4 + G_5 \end{bmatrix} \begin{bmatrix} V_1 \\ V_2 \\ V_3 \end{bmatrix}.$$

This requires the conductance matrix to be subsequently modified as follows:

$$\begin{bmatrix} 0 \\ 0 \\ -I_2 \end{bmatrix} = \begin{bmatrix} G_1 + G_2 & -G_2 - G & 0 \\ -G_2 & G_2 + G_3 + G_4 & -G_4 \\ 0 & -G_4 & G_4 + G_5 \end{bmatrix} \begin{bmatrix} V_1 \\ V_2 \\ V_3 \end{bmatrix}.$$

Once in this form the matrix equation can be solved using program filename 'NA'.

⟩ 2.5 Mesh or node analysis for circuits with voltage and current sources

Using the property that $R = 1/G$ and that voltage and current generators can be made equivalent, the circuits in figures 2.2 and 2.5 can both be redrawn and analysed by the alternative method.

Where sources occur as pure elements, without appropriately associated resistances, making it impossible to produce voltage or current equivalents, the technique shown below for combining two meshes or two nodes can be employed.

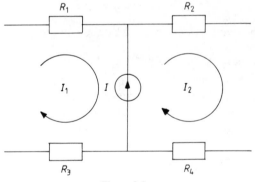

Figure 2.9

Figure 2.9 represents part of a circuit labelled for mesh analysis. The current source which is common to the two meshes shown cannot be replaced by a voltage equivalent and the voltage across it is indeterminate. The two mesh currents I_1 and I_2 are, however, related by the equation

$$I_2 = I_1 + I$$

where I is known. Only one equation is therefore required for the solution of both mesh currents. To provide the necessary equation the two meshes can be treated as one mesh by omitting the current source. This produces a mesh consisting of the outer parts of the two original meshes. The two mesh currents are used in the appropriate parts of the combined mesh to produce the voltage equation.

Similar reasoning can be used when performing node analysis on the circuit of which a part is shown in figure 2.10.

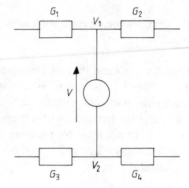

Figure 2.10

The two node voltages are related by the equation

$$V_2 = V_1 - V$$

where V is known, hence only one equation is required for the solution of the two values. This can be obtained by treating the two nodes as a single node in which the voltage source is included as an internal link. The need to include the indeterminate voltage source current is thus removed since it contributes nothing to the external current flow to or from the combined node.

The two voltages, V_1 and V_2, must be used as appropriate to calculate the current to or from the combined node.

⟩ **2.6 Norton and Thevenin equivalent circuits**

2.6.1 Introduction
When attention is to be focused on one part of a more complex circuit, Norton and Thevenin equivalent circuits can be used. These equivalents

replace the whole of the circuit with the exception of the part under consideration by a single current or voltage generator consisting of a source and appropriate internal resistance. A common application of this principle is in expressing the equivalent circuit of an amplifier or oscillator at the output terminals. In these applications the system circuitry is not the important consideration but rather the response in a load connected to the output.

Figure 2.11 shows the two equivalents related to a particular circuit used to demonstrate the principles.

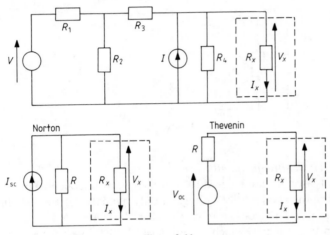

Figure 2.11

2.6.2 Norton equivalent

For a current generator the total source current is only available at the terminals under short-circuit conditions. It follows therefore that for equivalence the current source must be equal to the current, I_{sc}, which would flow in a short circuit replacing the circuit branch being considered, as shown in figure 2.12.

The value of the internal resistance of the generator, R, can be calculated using the principle of current division given in §1.7.2. Thus

$$I_x = I_{sc}R/(R + R_x)$$

$$R = R_x I_x/(I_{sc} - I_x).$$

To determine I_x and I_{sc} requires two mesh analyses of the circuit, first with R_x present and then with it replaced by a short circuit. Programs under filename 'MA' can be used for this purpose.

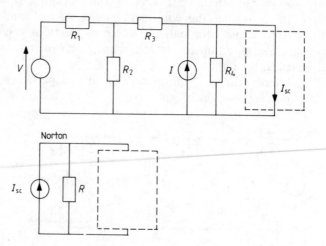

Figure 2.12

2.6.3 Thevenin equivalent

Similar reasoning can be used for the Thevenin equivalent. For a voltage generator the total source voltage is only available at the terminals on open circuit. This voltage must therefore be equal to V_{oc}, the voltage between the terminals which are produced in the circuit when the branch being considered is removed, as shown in figure 2.13.

The internal resistance of the generator, R, can be determined by using the principle of potential division given in §1.7.1. Thus

$$V_x = V_{oc} R_x / (R + R_x)$$

$$R = R_x (V_{oc} - V_x) / V_x.$$

V_x and V_{oc} can be determined by node analysis using programs under filename 'NA'.

Alternatively, the principle of generator equivalence, §1.8.4, can be used where

$$V_{oc} = R I_{sc}.$$

Hence either equivalent can be determined from the other and the one determined initially will depend on the circuit under consideration.

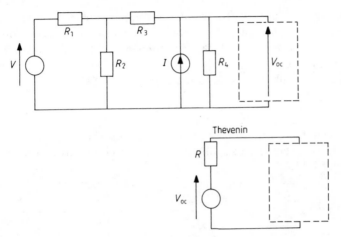

Figure 2.13

2.6.4 Alternative determination of the internal resistance, R

The significance of the internal resistance, R, in relation to the circuit and an alternative method for determining the value can be demonstrated by applying the principle of superposition to the circuits in figure 2.14.

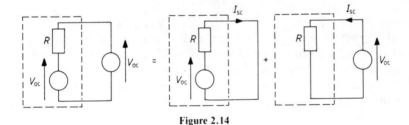

Figure 2.14

The circuit to be replaced by a current or voltage generator is unspecified and represented by the Thevenin equivalent shown within a dotted rectangle. This is a so-called 'black box' approach frequently used for reference to electronic systems without a detailed description of the circuitry. Since the voltage between the terminals on open circuit, i.e. with no current flowing, is V_{oc}, then connecting a voltage source of this value to the terminals still maintains zero current.

Applying superposition by first reducing the external voltage source to zero causes the short-circuit current, I_{sc}, to flow between the terminals.

This means that in order to produce a resultant current of zero with the external source present, a current of I_{sc} must flow in the opposite direction when the external source is applied and the internal sources are reduced to zero.

From this analysis it follows that, since

$$I_{sc} = V_{oc}/R$$

R must be the resistance between the terminals when all internal sources are reduced to zero. This is illustrated for the circuit given in figure 2.11 by figure 2.15.

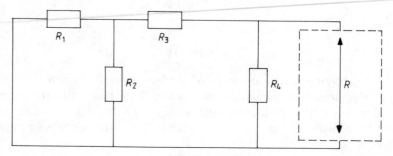

Figure 2.15

Program filename 'NT' illustrates the process for determining the Norton and Thevenin equivalents sequentially and then gives a numerical example. The element values in the example have been chosen to enable the parameter input to be performed by inspection. Users less familiar with the subject matter can use programs 'MA' and 'CC' to produce the required data.

2.6.5 Summary of Thevenin and Norton equivalents

The determination of the Thevenin and Norton equivalents for a particular circuit may be possible by analysis. In many instances the circuit complexity will render analysis inappropriate and direct measurement will be required to produce the equivalents.

The summary of the relationships given in figure 2.16 indicates the basis of the various analysis and measurement methods.

The problem-solving programs which can be used as appropriate are under filenames 'CC', 'MA' and 'NA'.

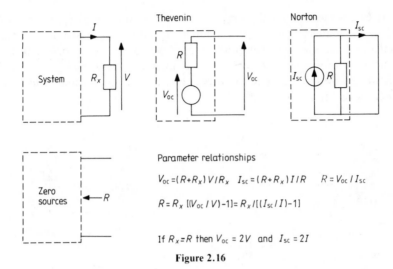

Figure 2.16

⟩ 2.7   Power transfer

In a number of situations it is desirable to maximize the transfer of power from source to sink, a typical example being an amplifier with a device requiring maximum power connected to its output terminals.

The conditions of power transfer from a generating system to a load can be investigated by reference to the Thevenin equivalent for the system shown in figure 2.17 together with the parameters for a specific example.

For the extremes of load resistance, short and open circuit, no power is transferred because the load voltage and current are zero, respectively. This suggests that somewhere between the extremes the power transferred will reach a maximum.

Power(P) = load voltage × load current

$$P = V_{oc}R_x/(R + R_x)V_{oc}/(R + R_x)$$
$$P = R_x[(V_{oc}/(R + R_x)]^2$$

Efficiency(E) = Power transferred/power generated × 100 %

$$E = P/[V_{oc}V_{oc}/(R + R_x)] × 100 \%$$
$$E = R_x/(R + R_x) × 100 \%.$$

The graphs of power transferred and efficiency against load resistance are shown in figure 2.17.

It can be seen from the graphs that maximum power is transferred when the load resistance, R_x, is equal to the internal resistance of the generator R. For this condition the efficiency is 50% since half the generated power is absorbed by the load and half by the internal resistance. Maximum efficiency occurs for any load value when the internal resistance is a minimum.

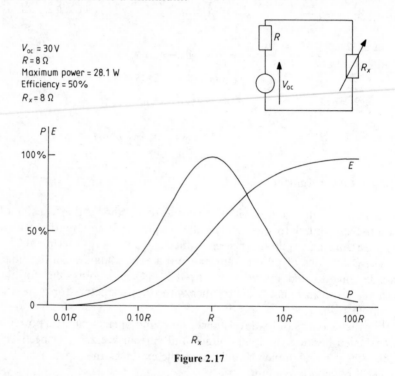

$V_{oc} = 30\,V$
$R = 8\,\Omega$
Maximum power = 28.1 W
Efficiency = 50%
$R_x = 8\,\Omega$

Figure 2.17

The criteria for maximum power transfer and maximum efficiency are clearly different and which is the most important consideration will depend on the application. Appendix A2.7.1 gives the traditional calculus analysis for determining the maximum value. Appendix A2.7.2 briefly describes the algorithm used for determining the maximum value by computer without using calculus.

Program filename 'MAXPOW' computes the maximum power and the load at which it occurs, using the above expressions on user-defined parameters. It then draws the graphs of power and efficiency against load resistance.

⟩ **2.8   Some common network configurations**

The circuits shown in figure 2.18 are examples of some common element configurations. These circuits are shown as two-port networks, with terminals 1 and 2 the input port and 3 and 4 the output port, a port being a pair of terminals where energy is imported to, or exported from, the circuit.

The networks in figures 2.18(a), (b) and (c) are π, T and lattice circuits, respectively. The lattice circuit is frequently drawn in the bridge form shown in figure 2.18(d). The bridge circuit has the particular property of producing zero output voltage if the four resistances obey the relationship

$$R_1/R_4 = R_2/R_3.$$

This property is demonstrated in practice problem 7.

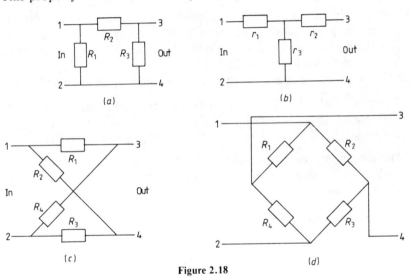

Figure 2.18

⟩ **2.9   Symmetrical and balanced networks**

Networks which appear electrically the same to any system connected to their input or output terminals are described as symmetrical. This is illustrated by the π, T and lattice circuits in figure 2.19.

The lattice network has the further property of being balanced since it would appear the same if terminals 2 and 4 were interchanged with terminals 1 and 3. A balanced network cannot have a common input and output terminal like the example of terminals, 2 and 4, for the π and T circuits.

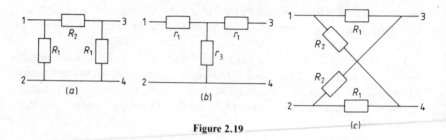

Figure 2.19

⟩ 2.10 π and T equivalence

π and T circuits can be made equivalent as far as the terminations are concerned provided the resistance values satisfy two sets of three equations. The equations are produced by considering the resistance between any pair of terminals.

Referring to the circuits in figure 2.18(a) and (b), these are as follows:

Terminals 1 and 2 $\qquad r_1 + r_3 = [R_1(R_2 + R_3)]/(R_1 + R_2 + R_3)$
Terminals 1 and 3 $\qquad r_1 + r_2 = [R_2(R_1 + R_3)]/(R_1 + R_2 + R_3)$
Terminals 3 and 4 $\qquad r_2 + r_3 = [R_3(R_1 + R_2)]/(R_1 + R_2 + R_3).$

These equations can be rearranged algebraically to make either set the subject and produce the six equations shown in table 2.1. The appropriate set can then be used for the conversion from π to T or T to π.

Table 2.1 π and T equivalents.

π to T	T to π
$r_1 = R_1R_2/(R_1 + R_2 + R_3)$	$R_1 = (r_1r_2 + r_2r_3 + r_3r_1)/r_2$
$r_2 = R_2R_3/(R_1 + R_2 + R_3)$	$R_2 = (r_1r_2 + r_2r_3 + r_3r_1)/r_3$
$r_3 = R_3R_1/(R_1 + R_2 + R_3)$	$R_3 = (r_1r_2 + r_2r_3 + r_3r_1)/r_1$

Program filename 'PT' solves these equations for conversion in either direction and can be used with the appropriate practice and other problems. A numerical example with the computer printout is given in figure 2.20.

Circuits which are similar to the π and T electrically but have terminals 2 and 4 brought together to produce three terminal structures are referred to as delta and star (sometimes Y) networks respectively. These three terminal forms are the circuit structures used in three-phase systems covered in §4.7. The relationships given in table 2.1 can be used for delta to star conversion.

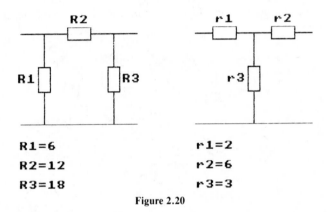

R1=6
R2=12
R3=18

r1=2
r2=6
r3=3

Figure 2.20

) 2.11 Iterative networks

An iterative network is one for which the input resistance is equal to the resistance connected to the output terminals, as shown in figure 2.21 for a symmetrical T circuit.

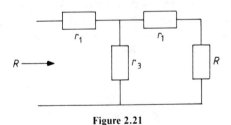

Figure 2.21

Rearranging the expression for the iterative resistance R produces the equation

$$r_3 = (R^2 - r_1^2)/2r_1.$$

The equation indicates that there is not a unique solution but that r_1 must be less than R to produce a positive value for r_3.

Figure 2.22 shows a numerical example which can be verified using program filename 'CC'.

The iterative condition for other circuit structures can be derived in a similar manner. The significance of such networks is that a number can be connected sequentially, cascaded, between a generator and a load while maintaining the same terminal conditions for the generator and each network.

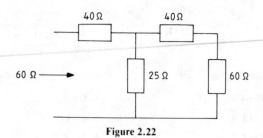

Figure 2.22

⟩ 2.12   Attenuators

2.12.1   Attenuation

Attenuation refers to the reduction of a signal which results from transmission through a resistive medium. Amplification is an opposite effect and is necessary not only to overcome attenuation due to transmission losses but also to increase a signal to the required level for a specific purpose. Variable amplification of electronic signals is usually achieved by fixed amplification coupled with variable attenuation. The sliding contact resistance shown in figure 1.26 is a common example of the mechanism for obtaining the variation.

2.12.2   Attenuators

For some applications a fixed and known value of attenuation may be required and can be achieved by using the type of network shown in figure 2.23.

When a passive network is connected between a generator and a load the attenuation which occurs is referred to as the insertion loss. The loss is frequently quoted in decibels, dB ($20 \log A$), where A is the ratio of the voltages across the load without and with the network present.

For the simplified case of the symmetrical iterative T network shown in figure 2.23,

$$A = V_1/V_2 = (R + r_1)/(R - r_1).$$

The insertion loss for other circuits can be derived in a similar manner.

Cascaded attenuators have an overall attenuation equal to the sum of the individual attenuation values in dB. Using iterative networks enables the attenuation to be predetermined since the cascaded networks operate with a standard termination.

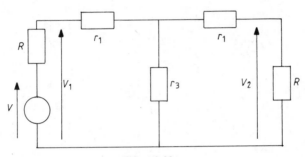

Figure 2.23

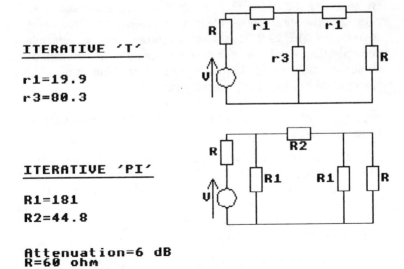

ITERATIVE 'T'

r1=19.9
r3=80.3

ITERATIVE 'PI'

R1=181
R2=44.8

Attenuation=6 dB
R=60 ohm

Figure 2.24

Program filename 'ATT' computes the symmetrical iterative π and T networks for a given attenuation and load and the printout for a numerical example is given in figure 2.24.

⟩ 2.13   Practice problems

1   Use mesh analysis to determine the current and power in each of the resistances in the circuit shown in figure 2.25.

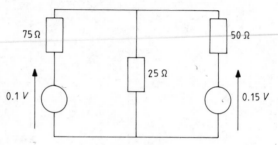

Figure 2.25

2   By analysing the circuit in figure 2.25 with each voltage source in turn demonstrate the principle of superposition. Use the values of current for the 25 Ω resistance to show that the principle of superposition is *not* valid for power calculation, i.e. the sum of the values of power obtained by each source acting alone is not equal to the total power with both sources operating.

3   Figure 2.26 shows a π circuit connected to two voltage generators. Determine the following:

   (i)   the current and power in each resistance of the π circuit;
  (ii)   the Thevenin equivalent for the remainder of the circuit, regarding the 10 Ω resistance as a load connected to it;
 (iii)   the new value for the 10 Ω resistance required for it to draw maximum power from the circuit; and
 (iv)   the equivalent T circuit for the π circuit.

    By repeating the mesh analysis with the equivalent T network, show that the source currents are equal to those with the π circuit

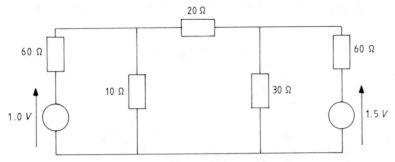

Figure 2.26

present and that the same total power is absorbed by the π and T circuits.

4   For the circuit shown in figure 2.26 replace the voltage generators with equivalent current generators and then determine the voltage across the 20 Ω resistance using node analysis.

5   Determine the current in each element of the circuit shown in figure 2.27.

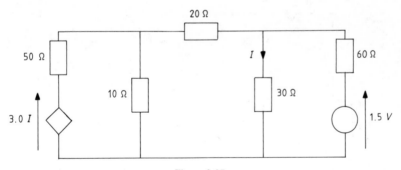

Figure 2.27

6   The circuit shown in figure 2.28 is a resistance net conduction field analogue. Perform

(i)   a mesh analysis on the circuit to determine the current source required to produce the same node voltages, and

(ii) a node analysis to determine the voltage of each node.

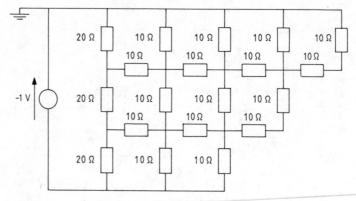

Figure 2.28

7 The circuit shown in figure 2.29 represents a resistance strain gauge bridge which in this context can be regarded as a bridge circuit with four variable resistances. Two of the four resistances, labelled r_1, increase and two, labelled r_2, decrease.

By using node analysis plot a graph of V_{ba} against percentage change in resistance for five '1%' changes in the bridge resistances. The initial value of the resistances r_1 and r_2 is 100 Ω.

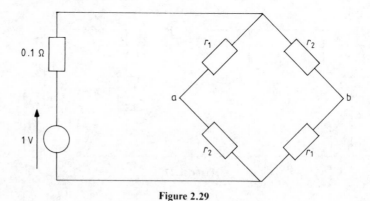

Figure 2.29

8 Determine R for the circuit shown in figure 2.30.

9 The bridged T circuit shown in figure 2.31(a) and the parallel T circuit shown in figure 2.31(b) are inserted between a voltage

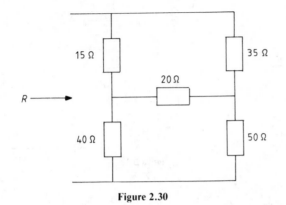

Figure 2.30

Figure 2.31

generator with an internal resistance of 60 Ω and a load resistance of 60 Ω. Determine the insertion loss in dB in both cases.

10 The circuit shown in figure 2.32 consists of two symmetrical iterative π attenuators cascaded between a voltage generator and a load resistance.

   (i)   Determine the values of R_1, R_2, R_3 and R_4 for the π attenuators to produce 10 and 30 dB of attenuation respectively.
   (ii) Replace the π networks with equivalent T networks.
   (iii) Use mesh analysis on the circuit produced in (ii) to confirm that the overall loss is 40 dB.
   (iv) Determine the Thevenin equivalent of the circuit for the 50 Ω load resistance.

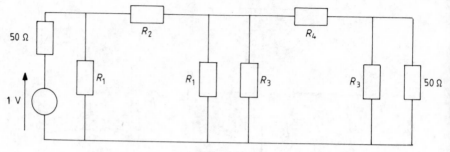

Figure 2.32

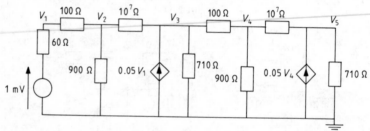

Figure 2.33

11   The circuit shown in figure 2.33 represents the equivalent circuit for a two-stage amplifier. Determine the node voltages V_1, V_2, V_3, V_4 and V_5 and the stage and overall voltage gains V_3/V_1, V_5/V_3, V_5/V_1 and $V_5/1$ mV.

This circuit is only valid when capacitance effects are negligible. Analysis techniques necessary to solve problems involving capacitance and a range of frequencies are given in later chapters.

〉 Chapter 3

〉 Sine Waves and Basic Single-phase Theory

〉 3.1 Introduction

Circuit responses to sinusoidal variations of voltage and current with time have a central significance in circuit theory and analysis. The sine wave is important, both as a source waveform and an analysis tool.

Probably the most notable examples of sine-wave sources are electricity supply systems. These distribute large amounts of energy through national grid circuits to domestic and commercial consumers by providing them on their premises with a sinusoidal voltage supply operating at a specified voltage and frequency. Much lower power sources widely exist in electronic equipment in the form of sine-wave oscillators as fixed or variable voltage and frequency devices.

From an analytical viewpoint the fact that any repetitive waveform can be synthesized by summing a series of frequency related sine waves provides an approach to analysing the response of circuits to non-sinusoidal signals. This is achieved using the principle of superposition with a series of pure sinusoids. By this technique the response of an unlimited range of waveforms can be determined using the principles of single sine-wave analysis.

The purpose of this chapter is to introduce some of the fundamental properties of sinusoidal responses, to provide the foundation and appreciation from which to progress to the more analytical techniques of Chapter 4.

〉 3.2   Sine-wave parameters

3.2.1   Amplitude and frequency

The instantaneous value of a voltage which varies sinusoidally with time is given by the equation

$$v = V_m \sin (\omega t).$$

The corresponding graph of voltage against time is shown in figure 3.1.

Since sine is an angle function, the sinusoidal variation with time is produced by a time-varying angle ωt, where t is the time in seconds and ω the rate of angle variation in radians/second. V_m is the maximum value or amplitude.

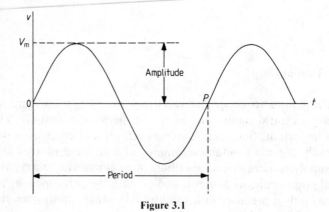

Figure 3.1

The time for one cycle is the period, P s, after which time the wave repeats itself. The repetition frequency, f, is therefore $1/P$ cycles/second (hertz).

The unit of angle is the radian (rad). The unit of frequency is the hertz (Hz). The sine-wave parameters are related as follows:

$\omega P = 2\pi$

$P = 2\pi/\omega$ $P = 1/f$ s

$f = 1/P$ $f = \omega/(2\pi)$ Hz

$\omega = 2\pi f$ $\omega = 2\pi/P$ rad s^{-1}.

3.2.2   Phase and phase difference

The phase of a sine wave refers to its position in time and therefore requires a time reference. Sine waves which are displaced in time have a phase difference relative to a common time reference. The instantaneous values of the two waves of equal amplitude shown in figure 3.2 are given by

$$v_1 = V_m \sin(\omega t + A) \qquad v_2 = V_m \sin(\omega t + B)$$

where A and B are constant and referred to as phase angles.

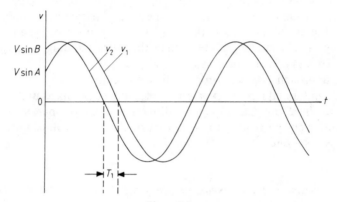

Figure 3.2

The phase difference of two sine waves which are displaced in time is usually expressed as an angle. Expressed in angle form, phase difference is frequency independent for a given relative position of two waves.

For example, two sine waves which have a phase difference of $\frac{1}{2}\pi$ rad are always one quarter of a cycle displaced relative to each other whatever the frequency. In terms of time, two sine waves which have a phase difference of $\frac{1}{2}\pi$ rad are displaced by varying amounts of time at different frequencies, i.e. 20 ms at 50 Hz, 10 ms at 100 Hz, etc.

The phase difference between two sine waves is determined by subtracting their instantaneous angles at any instant in time. For the waves shown in figure 3.2 the phase difference is given by

$$(\omega t + A) - (\omega t + B) = (A - B).$$

If $A > B$, v_1 is said to *lead* v_2; if $A < B$, v_1 *lags* v_2.

The relationship between phase difference expressed as an angle and the time displacement, T_1, between the two waves is

$$T_1 = P(A - B)/(2\pi) \, \text{s}.$$

Alternative angle unit

In some respects the radian is not the most convenient unit of angle and it is common practice to use the degree as an alternative unit for phase angles, where

$$1 \text{ degree} = 2\pi/360 \text{ rad.}$$

The approach adopted here is to use either unit for specifying phase angles or phase differences. Where the phase angle, A, is part of a time-varying angle, $(\omega t + A)$, it is specified separately in radians or degrees but as radians only in the total angle, e.g. if the phase angle A is given by $A = \frac{1}{2}\pi$ rad or $A = 90°$, then the corresponding time-varying angle, $(\omega t + A)$, is specified as $(\omega t + \frac{1}{2}\pi)$ rad.

This approach is a compromise between convenience and common practice and avoidance of expressing total angles in mixed units.

Where phase angles and/or phase differences are input parameters to a program, then the required unit is part of the data prompt unless it is otherwise obvious.

3.2.3   *Average and root-mean-square values*

Average value

The average value V_{av} of a parameter $v(t)$ which varies with time t is defined by the equation

$$V_{av} = \left(\int_{t=0}^{t=P} v(t) \, dt \right) P^{-1}$$

where P is the time over which the average value is calculated. $V_{av}P$ is the area under the graph of $v(t)$ against t between $t = 0$ and $t = P$.

Since the sine wave is symmetrical, with identical positive and negative half cycles, the average value taken over one or any integer number of cycles is zero.

The half-cycle average value is used in some instances, notably with rectified waveforms where the negative half cycle is either removed or inverted.

The average value of a sinusoidal signal taken over half of one cycle is given in terms of maximum value by the equation

$$V_{av} = (2/\pi) V_m.$$

This relationship is derived in Appendix A3.2.3.

Root-mean-square value

As indicated in §1.5.4, the RMS value is literally described, being in this instance the square root of the mean value of the sinusoidal signal squared.

The RMS value can be determined by integration or alternatively deduced from the sine squared wave (vs) shown in figure 3.3.

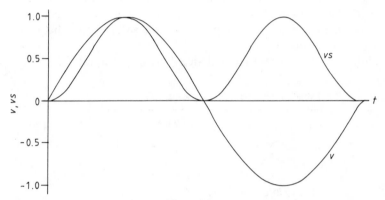

Figure 3.3

In keeping with the philosophy of the CIT, program filename 'INT' uses the numerical method described in Appendix A6.5 to determine the average and RMS values for a sine wave. Standard functional methods are equally appropriate, as shown in Appendix A3.2.3.

Figure 3.3 shows that the sine squared wave is itself a sine wave of twice the frequency of the wave from which it is derived, with the addition of a constant component. The result of squaring the sine wave can be expressed by the equation

$$(V_m \sin \omega t)^2 = \tfrac{1}{2}V_m^2 + \tfrac{1}{2}V_m^2 \sin (2\omega t + A).$$

The mean value of $(V_m \sin \omega t)^2$ over one cycle is therefore equal to $\tfrac{1}{2}V_m^2$ because the average value of $\tfrac{1}{2}V_m^2 \sin (2\omega t + A)$ over its two corresponding cycles is zero. Hence the RMS value of $V_m \sin \omega t$ is given by

$$V_{rms} = \sqrt{(\tfrac{1}{2}V_m^2)} = V_m/\sqrt{2}.$$

For the example shown in figure 3.3, produced by program filename 'RMS', $V_m = 1$ V and $A = -90°$.

Sinusoidal voltages and currents are usually specified by RMS values unless otherwise stated.

RMS values have particular significance in calculations of power, as described in §1.5.4 and discussed further in §3.8.

> **3.3   Reactance and susceptance**

As previously shown in figure 1.20, when inductance and capacitance are connected to sinusoidal sources the differential voltage–current relationship for these elements produces a $90°$ phase difference between the voltage and current waves.

This can be demonstrated analytically as shown in table 3.1.

Table 3.1   Reactance and susceptance.

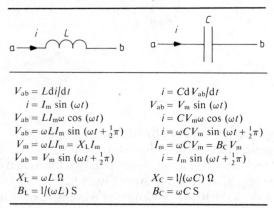

$V_{ab} = L \, di/dt$	$i = Cd \, V_{ab}/dt$
$i = I_m \sin(\omega t)$	$V_{ab} = V_m \sin(\omega t)$
$V_{ab} = LI_m \omega \cos(\omega t)$	$i = CV_m \omega \cos(\omega t)$
$V_{ab} = \omega LI_m \sin(\omega t + \frac{1}{2}\pi)$	$i = \omega CV_m \sin(\omega t + \frac{1}{2}\pi)$
$V_m = \omega LI_m = X_L I_m$	$I_m = \omega CV_m = B_C V_m$
$V_{ab} = V_m \sin(\omega t + \frac{1}{2}\pi)$	$i = I_m \sin(\omega t + \frac{1}{2}\pi)$
$X_L = \omega L \; \Omega$	$X_C = 1/(\omega C) \; \Omega$
$B_L = 1/(\omega L) \; S$	$B_C = \omega C \; S$

The ratio of the maximum values, voltage(V_m)/current(I_m) and the reciprocal ratio I_m/V_m are termed *reactance* and *susceptance* respectively.

The symbols used are X and B with the upper case subscripts 'L' and 'C' indicating the inductive and capacitive quantities, respectively.

From the definition it follows that

$$B = 1/X.$$

The unit of reactance is the ohm (Ω). The unit of susceptance is the siemen (S).

3.3.1   Frequency dependence

Graphs of inductive and capacitive reactance and susceptance for typical element values are given in figure 3.4 and show the frequency-dependent nature of these element properties. This dependence means that the current amplitude produced by a variable-frequency fixed-amplitude voltage source will thus decrease with frequency when the source is applied to an inductance and increase when it is applied to a capacitance.

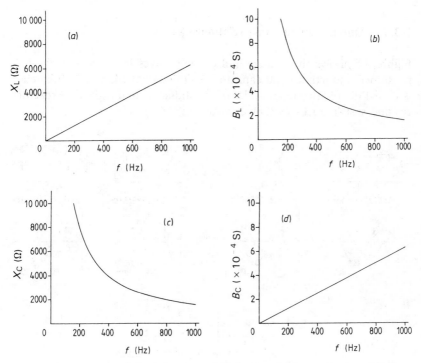

Figure 3.4 In (a) and (b), inductance = 1 H; in (c) and (d), capacitance = 0.1 μF.

Frequency-dependent properties of circuit elements are used in circuit design for separating signal components of different frequencies in a particular voltage or current waveform. This can be achieved by subjecting the different frequencies present to varying levels of attenuation and/or providing alternative conducting paths for specific signal frequencies and not others. Circuits designed for this purpose are broadly described as filters.

Capacitance is used in many electronic circuits for separating direct and alternating current components using the property that the capacitor presents an open circuit to direct current but a low reactance to high-frequency alternating current. Circuits can thus be designed with one conducting path for direct current and another for alternating current.

In general, inductance and capacitance are used separately or together in alternating current circuits for a wide range of filtering applications.

⟩ 3.4 Amplitude and phase relationships

Figure 3.5 shows the voltage and current waves in series and parallel, resistance, inductance and capacitance circuits. As the source frequency is varied the relative magnitudes of the inductive and capacitive reactances and susceptances vary proportionately.

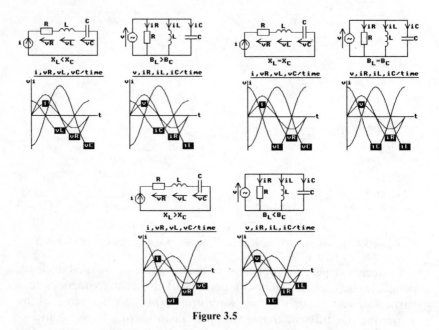

Figure 3.5

At a particular frequency the reactances and susceptances are equal. The frequency at which this occurs can be determined as follows:

$$X_L = X_C$$

therefore

$$\omega L = 1/\omega C$$

hence $\qquad\qquad\qquad \omega = (LC)^{-1/2} \text{ rad s}^{-1}$

and $\qquad\qquad\qquad f = 1/[2\pi(LC)^{1/2}] \text{Hz}.$

The graphs in figure 3.5 are given for three increasing source frequency conditions corresponding to frequencies below, equal to and above the value derived above. In detail these are as follows.

(a) $\omega < (LC)^{-1/2}$, the inductive reactance therefore being less than the capacitive reactance;

(b) $\omega = (LC)^{-1/2}$ making the two reactances equal, and

(c) $\omega > (LC)^{-1/2}$, the inductive reactance being greater than the capacitive reactance.

Series connection

For the series circuit the current is the same in all three components.

Charge does not actually flow through a capacitor since it consists of a pair of conductors insulated from each other. However, the effect of charge flowing to and from the conductors under alternating current conditions appears as continuous current external to the capacitor terminals. Since the current is common to all three elements it is taken as the phase reference. The voltages relative to the current are in phase for the resistance, $90°$ leading for the inductance and $90°$ lagging for the capacitance.

The amplitudes of the voltage waveforms are proportional to the resistance and the respective reactances at the source frequency. Thus for case (a) the amplitude of the inductance voltage is less than that for the capacitance; for case (b) the inductance and capacitance voltage amplitudes are equal and for case (c) the inductance voltage amplitude is greater than the capacitance voltage. The current source voltage is found by adding the three element voltages.

As shown in figure 3.5, the inductance and capacitance voltage waves are $180°$ out of phase. This has a particular significance when the two waves are equal as in case (b), because the sum of the voltages across the inductance and capacitance added together is zero. The voltage across the resistance is thus equal to the source voltage.

Parallel connection

For the parallel circuit the voltage is the same across each component and it is therefore used as the phase reference. The resistance current is in phase, the inductance current is $90°$ lagging and the capacitance current is $90°$ leading, relative to the circuit voltage. The amplitude of the current waveforms is proportional to the conductance and the respective susceptances at the source frequency. For case (*a*) the inductance current is therefore the greater of the two susceptance currents; in case (*b*) they are equal and in case (*c*) the inductive susceptance conducts the lower current of the two susceptances. The voltage source current is the sum of the three element currents.

As shown in figure 3.5 the inductance and capacitance current waves are 180° out of phase. When the two waves are equal, as in case (*b*), the inductance and capacitance currents cancel and the source current is thus equal to the resistance current.

General properties
For both circuits and more generally, the models for resistance and conductance are frequency independent. In reality resistors and conductors can depart from this ideal model particularly at high frequencies.

The frequency at which the reactances and susceptances are equal produces a resonant condition in both circuits. For this condition the effects of the inductance and capacitance cancel as far as the source is concerned, to which the circuit appears as a pure resistance. Within the resonant circuit the inductance and capacitance voltages and currents can separately be very significant. Resonant conditions are discussed in more detail in Chapter 4.

Program filename 'AP' produces the waveform displays in figure 3.5 for the three separate conditions. The waves are displayed in sequence to highlight the amplitude and phase relationships and the principles involved.

⟩ **3.5   Types of response**

Introduction
When the condition of a system is changed, e.g. if it is energized from an unenergized state, its sequence of responses is typically as shown in figure 3.6.

The interpretation put on the terms applied to circuit responses is frequently related to the analytical method by which they are determined. It is necessary therefore, since this is a computer treatment, to define the terms in this context. The meaning attached to the terms initial, steady-state and transient response is literal and more precisely as defined below.

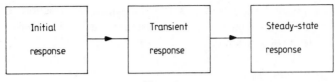

Figure 3.6

(i) *Initial* response refers to the set of conditions which exist for an instant immediately after the operating conditions of a system are changed.

(ii) *Steady-state* response describes the state of a system which is not changing. For a DC circuit this would typically be a constant voltage or current and for an AC system a repetitive waveform, i.e. with constant amplitude and frequency.

(iii) *Transient*, or changing, response refers to the state of a system as it changes from its initial to its steady-state response.

The transition from initial to steady-state response is continuous and there is no definite time when the transient response becomes the steady state. In general a system model will either have no transient response or one which persists indefinitely and approaches the steady-state response asymptotically.

For an actual system, the steady state will be considered to exist when there is no perceptible change in the response. This state is reached with varying degrees of rapidity depending on the system parameters.

Initial responses

The initial response of a circuit is determined by the voltage–current relationships for the circuit elements. The model for resistance permits instantaneous change of current and voltage; for inductance and capacitance, the current and voltage, respectively, cannot change instantaneously. The initial response of a circuit therefore, has all inductance currents and all capacitance voltages unchanged from their values immediately prior to the change in the circuit conditions. All other circuit parameters must then be compatible with these requirements.

Steady-state responses

Steady-state responses can be determined by various techniques. Frequency-domain analysis determines steady-state responses for systems with sinusoidal sources, expressing the response in terms of amplitude and phase data at a single frequency or over a range of frequencies.

Time-domain analysis expresses system responses as functions of time. Techniques associated with both of these approaches are considered in some detail in the remainder of this book.

Transient responses
Transient responses by their nature change with time and are therefore the subject of time-domain analysis with circuit currents and voltages being expressed as functions of time. This type of analysis is often required to investigate the possible existence of values atypical of the steady-state response and, if such values do exist, to ascertain how long they persist.

   Electronic and other physical systems are frequently subject to much greater stress when the system is suddenly changed in some manner than when it is operating under steady conditions. Typical examples of such changes in electrical systems are the instances when current flow is initiated or interrupted.

Steady-state sinusoidal waveforms
The waveforms shown in figure 3.5 are steady-state responses for which the time datum for the display is arbitrary and must not be confused with zero time. The waveforms shown of the capacitance voltage in the series circuit and the inductance current in the parallel circuit are not compatible with zero time at the time datum of the display. Assuming the elements were previously unenergized, an instantaneous change would be required to bring them to the indicated values. This would violate the properties of these elements.

   All the analysis in this chapter is concerned with steady-state responses and the point made above concerning the time datum for steady-state waveform displays is generally applicable.

⟩ **3.6   Sine-wave addition**

Section 3.4 has demonstrated the need to add sine waves of the same frequency to determine total voltages and currents in series and parallel circuits. Figure 3.7 shows the result of adding two sine waves of the same frequency. The resultant is a third sine wave of unchanged frequency with a magnitude and phase which depends on the magnitude and relative phases of the two waves added. The sum of the two sine waves can be expressed by the equation

$$V_1 \sin (\omega t + A) + V_2 \sin (\omega t + B) = V_3 \sin (\omega t + C).$$

Program filename 'AS' performs the addition of two sine waves with specified amplitude and phase. It can be used to demonstrate the process of sine-wave addition as well as for solving circuit problems as described below.

Sine waves which are in phase produce a resultant of amplitude equal to the sum of the amplitudes of the component waves. For waves not in phase the addition is analytically, though not computationally, more complex.

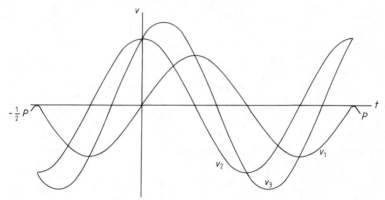

Figure 3.7 $V_1 = 3$, $V_2 = 4$, $V_3 = 5$. $A = 0_{rad(0deg)}$, $B = 1.57_{rad(90deg)}$, $C = 0.927_{rad(53.1deg)}$.

⟩ 3.7   Impedance and admittance

The circuits shown in figures 3.8 and 3.9 are series and parallel alternating current circuits with corresponding voltage and current waveforms. The combinations of resistance and/or conductance with reactance and/or susceptance are referred to as *impedances* or *admittances*.

The impedance, Z, is defined for a circuit by the relationship

$Z = $ (amplitude of circuit voltage)/(amplitude of circuit current).

The admittance, Y, is the reciprocal of impedance, hence $Y = 1/Z$ and is thus defined by

$Y = $ (amplitude of circuit current)/(amplitude of circuit voltage).

The unit of impedance is the ohm (Ω). The unit of admittance is the siemen (S).

3.7.1 Series circuits

The current for the elements connected in series in the circuits shown in figure 3.8 is the same in each element and therefore used as the phase reference.

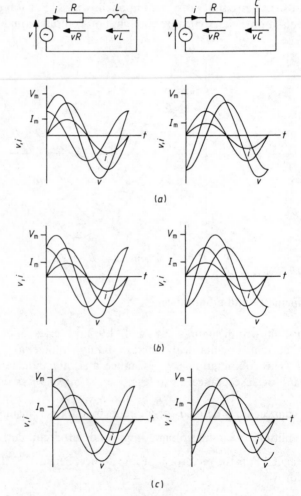

(a)

(b)

(c)

Figure 3.8

The circuit conditions are shown for three increasing frequencies where

(a) $X_L < R$ and $X_C > R$ making $v_L < v_R$ and $v_C > v_R$,
(b) $X_L = R$ and $X_C = R$ which produces equal voltages, and
(c) $X_L > R$ and $X_C < R$ with $v_L > v_R$ and $v_C < v_R$.

The phase difference between the source voltage and current in each case is dependent on the relative magnitudes of the element voltages. This in turn depends on the relative magnitudes of the resistance and reactance at the particular source frequency.

For case (b) where the resistance is equal to the reactance in both circuits a phase difference of $45°$ is produced. In the other examples the phase difference has a value between 0 and $90°$ dependent on the frequency.

For the inductive circuit the current lags the voltage, for the capacitive circuit it leads.

Program filename 'SC' produces the graphs shown in figure 3.8 in sequence and should be used to demonstrate the various magnitude and phase relationships and the underlying principles involved.

3.7.2 Parallel circuits
The parallel circuits shown in figure 3.9 are modelled in terms of conductance and susceptance to use the full range of models available. Resistance and reactance models would be equally valid and either set are used as convenience dictates. For the parallel circuit the voltage is the same for each of the elements and it is therefore used for the phase reference.

As for the series circuit, the conditions are shown for three increasing frequencies expressed in terms of the relative magnitude of the conductance and susceptance where

(a) $B_L > G$ and $B_C < G$ making $i_L > i_G$ and $i_C < i_G$,
(b) $B_L = G$ and $B_C = G$ which produces equal currents, and
(c) $B_L < G$ and $B_C > G$ with $i_L < i_G$ and $i_C > i_G$.

The phase difference between the source voltage and current in each case is dependent on the relative magnitudes of the element currents. This in turn depends on the relative magnitudes of the conductance and susceptance at the particular source frequency.

For case (b) where the conductance is equal to the susceptance in both circuits a phase difference of $45°$ is produced. In the other examples the phase difference has a value between 0 and $90°$ dependent on the frequency.

For the inductive circuit the current lags the voltage, for the capacitive circuit it leads.

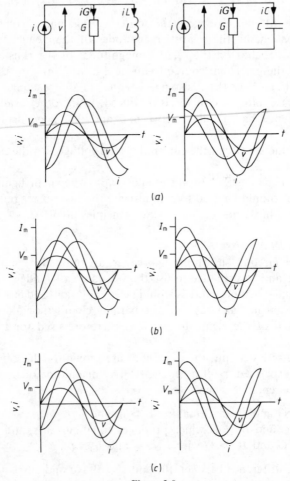

Figure 3.9

Program filename 'PC' produces the graphs shown in figure 3.9 in sequence and should be used to demonstrate the various magnitude and phase relationships and the underlying principles involved.

3.7.3 Reactance, susceptance, impedance and admittance models

The models for reactance, susceptance, impedance and admittance described in the above sections are elementary ones. Being only ratios of magnitude, they do not include information regarding the phase relationships of the voltages and currents involved. These are adequate models for a broad appraisal of the way circuits behave with sinusoidal supplies. For detailed analysis of alternating current circuits, however, they are not adequate and more sophisticated circuit models are required.

Models of reactance, susceptance, impedance and admittance which include phase-angle information are discussed in Chapter 4.

⟩ 3.8 Power

3.8.1 Instantaneous power

With sinusoidal supplies the voltage and current are continuously varying and the rate at which energy is being supplied or absorbed by a circuit is the instantaneous power, which likewise is a continually varying quantity. For the circuit shown in one-port 'black box' form in figure 3.10 the instantaneous power, p, is given by

$$p = V_{ab}i$$

where V_{ab} and i are the instantaneous values of voltage and current.

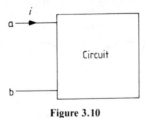

Figure 3.10

As stated in §1.3.5, if V_{ab} and i are both positive then p is positive and the circuit shown acts as a sink and absorbs energy. Alternatively, if V_{ab} is positive and i becomes negative then p is negative and the circuit acts as a source and supplies energy.

If the circuit consists solely of passive elements it can only act as a source if it has stored energy and therefore contains inductance and/or capacitance.

For sinusoidal supplies where there is a phase difference between the current and voltage, the sign of p alternates as the passive or active devices act as both source and sink during different parts of a cycle. Figure 3.11 shows voltage and current waves for the circuit in figure 3.10 and the resulting power wave produced by multiplying them together.

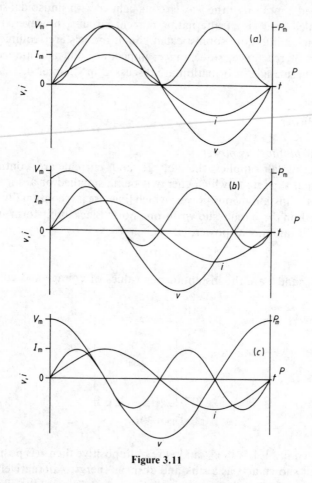

Figure 3.11

The instantaneous power is given by the equation

$$p = V_m \sin (\omega t + A) I_m \sin (\omega t + B) \text{ W}.$$

The three examples given are for voltage and current waves of the same frequency with the following phase relationships:

(a) the current and voltage in phase ($A = 0$, $B = 0$),
(b) the current lagging the voltage by $60°$ ($A = 60°$, $B = 0$),
(c) the current lagging the voltage by $90°$ ($A = 90°$, $B = 0$).

Similar waveforms are produced with leading phase angles and can be similarly displayed using program filename 'PW'.

The same values of the amplitude for the voltage and current waves are used in each example and the variations in the resulting power wave are due entirely to the different values of phase difference between them.

3.8.2 Power wave properties

The following properties of the power wave can be seen by reference to the three examples shown in figure 3.11.

1   The power wave has sinusoidal and steady value components.
2   The frequency of the sinusoidal component in the power wave is twice that of the frequency of the voltage and current waves.
3   The value of the steady component of power varies as the phase difference between the voltage and current varies and is zero when the phase difference is $90°$.
4   For given values of voltage and current amplitude, V_m and I_m respectively, the power-wave amplitude is the same for different values of phase difference. The power-wave amplitude is given by

$$P_m = \tfrac{1}{2} V_m I_m.$$

5   Combining the above observations it can be shown that the power wave is given by the expression

$$p = P_{av} + P_m \sin (2\omega t + C)$$

where P_{av} is the average value of power.

3.8.3 Average power

Since the sine wave is symmetrical, the average value of power is the mean of the maximum and minimum values of the power wave. This is equal to the amount the axis of symmetry of the power wave is offset along the power axis, in other words the steady component of the power expression described in §3.8.2.

Figure 3.11 shows that when the phase difference between the voltage and current is zero the power wave is always positive. At other values of phase difference, part of the power-wave cycle is negative as the sign of the voltage and current change relative to each other, producing a negative product.

Program filename 'PW' multiplies voltage and current sinusoids of the same frequency to produce the power waveform. The program also determines the maximum and minimum values of power and prints out the average value and power factor.

The waveforms in figure 3.11 can be demonstrated using program filename 'PW' with examples of voltage and current amplitude of 10 V and 1 A, respectively, to facilitate the interpretation.

3.8.4　Power factor

When the phase difference between voltage and current is zero the average power is given by

$$P_{av} = \tfrac{1}{2} V_m I_m = V_{rms} I_{rms}.$$

The RMS values arise because the average power is derived from the product of two in-phase sine waves which produces a $(\text{sine})^2$ function. $V_{rms} I_{rms}$ is the maximum value of average power which can be produced by a voltage and current wave. For phase differences greater than zero the average power is less, as shown by the waveforms in figure 3.11.

The ratio $P_{av}/V_{rms} I_{rms}$ is termed the power factor and the product $V_{rms} I_{rms}$ is frequently referred to as the volt–ampere or VA rating of the device, circuit or system being considered.

For supply systems operating at a constant voltage amplitude with a

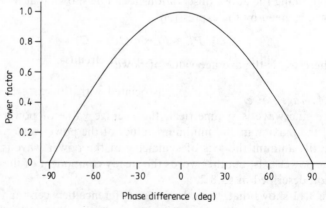

Figure 3.12

specified current capacity, the power and hence energy which can be delivered in a given time is a maximum when the power factor is unity. In some instances where equipment operates at lower power factors it is economically advantageous to improve the power factor to increase supply system efficiency. The subject of power factor improvement is covered in more detail in §4.6.3.

Figure 3.12 shows a graph of power factor against phase difference. Analysis shows that the graph is a cosine curve. In this instance it is produced by program filename 'PFG' using a similar algorithm to program filename 'PW' which multiplies two sine waves and then determines the average value. The cosine relationship is developed analytically in §4.6.1.

3.8.5 Power in passive components

The special cases of power dissipated in passive components by sinusoidal sources relate to the more general statements in §1.5.4.

For inductance and capacitance the average power is zero. This correlates in the case of sinusoidal signals with the $90°$ phase difference between voltage and current and corresponding zero power factor for these elements.

Resistance produces a zero phase difference between current and voltage and hence a unity power factor. In the terms described above the average power is therefore given by

$$P_{av} = V_{rms}I_{rms} = V_{rms}^2/R = I_{rms}^2 R.$$

These are the same as the general expressions given in §1.5.4 which are not waveform specific.

⟩ 3.9 Analysis of simple alternating current circuits

The analysis and computing tools associated with this chapter can be used for solving relatively basic AC circuit problems with sine-wave sources, as described below.

(i) The demonstration programs filenames 'AP', 'SC' and 'PC' should be used to develop an understanding of amplitude and phase relationships in sine-wave circuits and the properties of reactance, susceptance, impedance and admittance, prior to any attempt at problem solving.

(ii) Impedances, admittances and phase differences can be determined for various configurations of circuit elements by assuming unity amplitude current or voltage sources as appropriate and then using program filename 'AS' to add the sine-wave responses.

Two examples are shown in figures 3.13 and 3.14. Analysis of the circuit in figure 3.13 produces the following results: $X_L = 1257\ \Omega$, $V_R = 1000\ \text{V}$, $V_L = 1257\ \text{V}$, $V_R + V_L = 1.61 \times 10^3\ \text{V}$, $Z = 1.61 \times 10^3\ \Omega$ and $Y = 6.2 \times 10^{-4}\ \text{S}$. The high voltages are produced by the unity current amplitude.

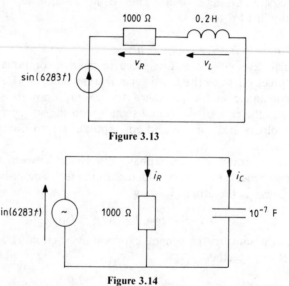

Figure 3.13

Figure 3.14

The phase difference between the source current and voltage is equal to $51.5°$ with the voltage leading.

For the circuit in figure 3.14 the circuit parameters are $B_C = 6.283 \times 10^{-4}\ \text{S}$, $I_R = 10^{-3}\ \text{A}$, $I_C = 6.283 \times 10^{-4}\ \text{A}$, $I_R + I_C = 1.183 \times 10^{-3}\ \text{A}$, $Y = 1.183 \times 10^{-3}\ \text{S}$ and $Z = 845.3\ \Omega$. The phase difference between the source current and voltage is equal to $32.1°$ with the current leading.

For the analysis of the two circuits in this section V_R, I_R, ..., etc, are maximum values.

(iii) Responses produced by given voltage or current sources can be found by first determining the impedance or admittance as described in (ii) above and then using the appropriate relationships from

$$I_m = V_m/Z \qquad \text{or} \qquad V_m = ZI_m$$

and

$$V_m = I_m/Y \qquad \text{or} \qquad I_m = YV_m.$$

Two examples are given in figures 3.15 and 3.16 which use the circuits from figures 3.13 and 3.14 for which the impedances and admittances have been previously determined in (ii).

The following results are obtained for the circuit in figure 3.15: circuit   current $= 1/Z = 6.21 \times 10^{-4}$ A, $V_R = 0.621$ V   and $V_L = 0.781$ V.

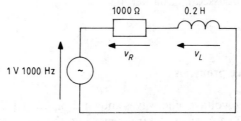

Figure 3.15

The following results are obtained for the circuit in figure 3.16: circuit   voltage $= 10^{-3} Z = 0.845$ V, $I_R = 8.45 \times 10^{-4}$ A   and $I_C = 5.31 \times 10^{-4}$ A.

For the analysis of the two circuits in this section V_R, I_R, ..., etc, are RMS values.

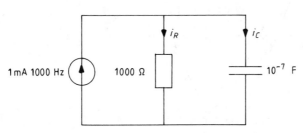

Figure 3.16

(iv) The average power and power factor of specific circuits can be computed by using program filename 'PW'. The voltage and current amplitudes and the phase difference must previously be determined by using the techniques in (i) and (ii) above.

For the examples given in figures 3.15 and 3.16 the values of power and power factor produced by program filename 'PW' are as follows.

For the circuit in figure 3.15, $P_{av} = 3.87 \times 10^{-4}$ W and power factor $= 0.623$; for the circuit in figure 3.16, $P_{av} = 7.16 \times 10^{-4}$ W and power factor $= 0.847$.

The techniques and principles covered in this chapter form the basis for the analysis of circuits with sinusoidal sources. The analysis of more complex circuits requires more sophisticated circuit models and more powerful analysis tools. Some of the more advanced models and processes are discussed in Chapter 4.

⟩ **3.10 Practice problems**

1 For the three voltage sine waves shown in figure 3.17 determine the amplitude of each wave, the frequency and the phase difference of v_1 relative to v_2.

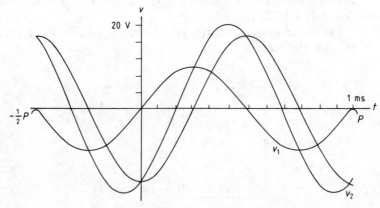

Figure 3.17

2 Demonstrate the following sine-wave additions using program filename 'AS'.

(i) $10 \sin \omega t + 10 \sin \omega t$

(ii) $30 \sin \omega t + 40 \sin \omega t$

(iii) $30 \sin \omega t + 40 \sin(\omega t + \pi)$

(iv) $100 \sin(\omega t + \frac{1}{6}\pi) + 100(\sin \omega t + \frac{1}{6}\pi)$

(v) $100 \sin(\omega t + \frac{1}{6}\pi) + 100 \sin(\omega t - \frac{2}{3}\pi)$

(vi) $10 \sin \omega t + 10 \sin(\omega t + \frac{1}{2}\pi)$

(vii) $10 \sin \omega t + 10 \sin(\omega t - \frac{1}{2}\pi)$

(viii) $240 \sin \omega t + 240 \sin(\omega t + \frac{1}{3}\pi)$

(ix) $240 \sin(\omega t - \frac{2}{3}\pi) + 240 \sin(\omega t - \frac{1}{3}\pi)$

(x) $240 \sin(\omega t + \frac{2}{3}\pi) + 240 \sin(\omega t - \pi)$.

The last three examples are significant in three-phase systems which are discussed in Chapter 4.

3 The series circuit shown in figure 3.18 is supplied by a sinusoidal voltage source of amplitude 1 V. Determine at frequencies of 100 Hz, 1000 Hz and 10 kHz the following:

(i)   the reactance and susceptance of the inductance and the capacitance;

(ii)   the impedance and admittance of the circuit;

(iii) the circuit current;

(iv) the voltage across each of the series elements;

(v)   the phase difference between the source voltage and current; and

(vi) the power and power factor of the circuit.

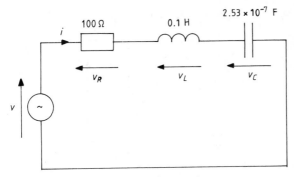

Figure 3.18

4 Figure 3.19 shows a parallel circuit supplied by a sinusoidal voltage source of amplitude 4 V. Determine at frequencies of 100 Hz, 1000 Hz and 10 kHz the following:

   (i)   the reactance and susceptance of the inductance and the capacitance;
   (ii) the impedance and admittance of the circuit;
   (iii) the circuit current;
   (iv) the current in each of the parallel elements;
   (v)   the phase difference between the source voltage and current; and
   (vi) the power and power factor of the circuit.

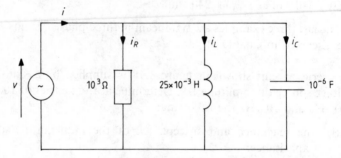

Figure 3.19

5 The circuit shown in figure 3.20 is used for coupling a signal to different parts of an electronic circuit. Determine the ratio of the voltage amplitudes V_2/V_1 and the phase difference between the two voltages at frequencies of 10 Hz, 1000 Hz and 10 kHz.

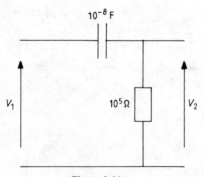

Figure 3.20

6 Figure 3.21 shows a circuit which is designed to separate the DC and AC components of current in a signal containing both types. The instantaneous current source is given by $i = (10^{-3} + 10^{-2} \sin \omega t)$ A. Use the principle of superposition with each of the components of current to determine the complete response.
Determine for values of ω of 63, 628 and 6283 rad s^{-1}:

   (i)   the frequency corresponding to each value of ω;
   (ii) the proportion of each current component which flows through each of the parallel paths;
   (iii) the voltage across the circuit at each value of ω; and
   (iv) the ratio of the alternating components of voltage across the circuit with and without the capacitance present.

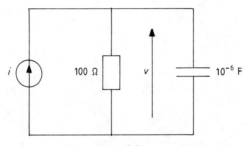

Figure 3.21

7 The circuit shown in figure 3.22 is designed to filter the alternating component in the source waveform to reduce the component of AC flowing in the 100 Ω resistance. The instantaneous voltage of the source which is derived from rectifying a 50 Hz supply is given by

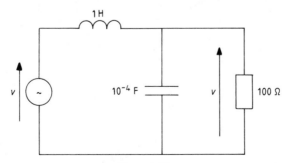

Figure 3.22

$v = [6.4 + 4.3 \sin (628t)]$ V. Use the principle of superposition with the source voltage components to determine the components of steady and alternating voltage across the 100 Ω resistance for

   (i)   the complete circuit;
  (ii)   the circuit with the capacitance reduced to zero; and
 (iii)   the resistance alone.

Calculate the relative effects of the filtering, in terms of the attenuation loss in db, to which the alternating component of voltage is subjected for conditions (i) and (ii) relative to (iii).

〉 Chapter 4

〉 Frequency-domain Analysis and Responses

〉 4.1 Phasors

The basic alternating current theory in Chapter 3 has demonstrated the need for more powerful models and analysis techniques which can be used to determine steady-state responses in circuits with sinusoidal sources. The process illustrated by program filename 'AS' of adding sine waves graphically is effective but quickly becomes too cumbersome for all but the most simple circuits.

Ways of modelling signals and circuit elements are required which encompass the three significant parameters of sine-wave analysis, namely, frequency, magnitude and phase. Well established analytical models exist, such as vectors and complex numbers, which are discussed briefly below.

The emphasis here is rather on using similar computer models to develop techniques which are appropriate to computer-based analysis while maintaining the essential properties of the established models.

It has been shown that in general in electronic circuits, sinusoidal signals undergo both magnitude and phase changes and that circuit properties such as impedance and admittance not only produce magnitude and phase changes but are also frequency dependent.

In introducing sine-wave signals it was stated that a process of waveform synthesis using single-frequency harmonic sinusoids, in conjunction with the principle of superposition, permits the analysis of circuits with an unlimited range of signal waveforms. It is only necessary therefore for models to contain magnitude and phase information at a

95

specific frequency even though both parameters are likely to be frequency dependent.

Circuit element models which contain magnitude and phase information are broadly referred to as *phasors*, although in specific terms they may be quite different. Vectors, complex numbers and the phasors described below for computer analysis are examples of three different types of model which are used for steady-state sine-wave analysis. All three can generally be regarded as phasors in this context, which employs their similar properties. They are referred to specifically in the following sections for identification.

4.1.1   Phasor properties

The application of Kirchhoff's laws to circuits with sine-wave sources requires the addition of sinusoidal voltages and currents. Except in resistance-only circuits, these signals will have phase differences for which the process of sine-wave addition is more complex than simply adding the amplitudes of the two waves.

Phasors must therefore not only carry magnitude and phase information but in addition require specially defined processes of combination.

The computer is very suited to manipulating defined quantities according to precisely defined processes and the problem-solving programs attached to this section of theory employ the definitions and processes set out below.

The term phasor is used here to mean a quantity with two distinct parts: a *magnitude* and an *angle*. The magnitude relates to the size of the property being modelled. The angle has a value between $-\pi$ and $+\pi$ radians or $-180°$ and $+180°$ and corresponds to the phase of, or phase difference between, the signal(s) associated with the model.

A voltage phasor is thus expressed as V_{radA} or V_{degB}, the unit of V being the volt and the unit of A and B being the radian or degree. The approach adopted to angle units for phasors is the same as that described in §3.2.2. The radian is used where it has a functional significance, otherwise the degree is used.

The basic property of a phasor can be illustrated by considering the voltage, $V\sin(\omega t + A)$, and the reference wave, $\sin \omega t$. The phasor relating these two sine waves is V_{radA}, V and A being the amplitude and phase of the time-varying voltage relative to the reference wave amplitude and phase, respectively.

Defining phasors in this way means that all sinusoidal voltages and currents in a system can be related to the same reference wave, $\sin \omega t$, by the appropriate phasors.

4.1.2 Phasor models

In addition to voltages and currents, circuit elements can be modelled as phasors. The passive elements are modelled in terms of resistance, conductance, reactance and susceptance phasors. Reactance and susceptance are necessary to model inductance and capacitance because the voltage–current relationships for these elements are frequency dependent. Combinations of resistance and/or conductance with reactance and/or susceptance are modelled as impedance or admittance phasors and can be treated as single elements.

Table 4.1 Phasor models.

Property	Phasor	Magnitude	Angle
Voltage	$V_{\text{rad} \pm A}$ $V_{\text{deg} \pm B}$	Amplitude or RMS value of voltage	Phase of V relative to reference
Current	$I_{\text{rad} \pm A}$ $I_{\text{deg} \pm B}$	Amplitude or RMS value of current	Phase of I relative to reference
Resistance	$R_{\text{rad}\,\theta}$ $R_{\text{deg}\,\theta}$	Amplitude of $V \div$ amplitude of I	Phase of V relative to phase of I
Conductance	$G_{\text{rad}\,\theta}$ $G_{\text{deg}\,\theta}$	Amplitude of $I \div$ amplitude of V	Phase of I relative to phase of V
Reactance	$X_{\text{rad} \pm \pi/2}$ $X_{\text{deg} \pm 90}$	Amplitude of $V \div$ amplitude of I	Phase of V relative to phase of I
Susceptance	$B_{\text{rad} \mp \pi/2}$ $B_{\text{deg} \mp 90}$	Amplitude of $I \div$ amplitude of V	Phase of I relative to phase of V
Impedance	$Z_{\text{rad} \pm A}$ $Z_{\text{deg} \pm B}$	Amplitude of $V \div$ amplitude of I	Phase of V relative to phase of I
Admittance	$Y_{\text{rad} \mp A}$ $Y_{\text{deg} \mp B}$	Amplitude of $I \div$ amplitude of V	Phase of I relative to phase of V

Circuit element phasors relate the voltage drop across the element to the current through it and hence in turn relate the phasors for these quantities. The magnitude of the phasor for each of the passive elements expresses the voltage–current amplitude ratio and the phasor angle is equal to the phase difference between the voltage drop across the element and the current through it. Since resistance and conductance produce no phase difference between current and voltage the phasors for these elements have zero angles. The sign of the angle in the other cases indicates the phase of the voltage drop relative to the current or vice versa depending on which parameter is taken as the phase reference.

Voltage and current sources are modelled in terms of either maximum or RMS values. The circuit voltage and current responses will likewise be maximum or RMS values. The angle of the source phasor indicates the source phase angle relative to the phase reference.

The phasors for the circuit elements and combination of elements are specified in table 4.1. The sign of the angles which apply to inductive and capacitive reactance and susceptance respectively are explained in §4.2.2.

Mutual inductance is modelled as a phasor in terms of reactance or susceptance in a similar manner to inductance. The sign of the phasor quantity in any given application will need to account for the sense of the magnetic coupling as described in §1.4.7.

Conventions

The conventions described for current and voltage in Chapter 1 are applied to phasors in a similar manner. Referring to figure 4.1 I is the phasor for the current in the direction shown and V_{ab} the phasor for the potential of point a relative to point b. $V_{ab} = V$ for the other conventions.

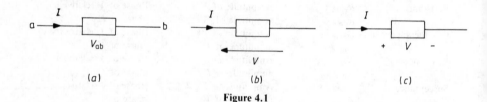

Figure 4.1

⟩ 4.2   Phasor processes

The power of phasor analysis derives from the special processes of phasor combination. These make it possible to determine circuit responses and element combinations in phasor form which can then be interpreted in terms of physical or standard circuit parameters.

The processes of phasor multiplication, division and addition are described below, followed by examples of their application.

4.2.1   Phasor multiplication and division

Multiplication

The process of phasor multiplication is defined by the equation

$$V_{\deg A_1} = (Z_{\deg A_2})(I_{\deg A_3})$$

where $V = ZI$ and $A_1 = A_2 + A_3$. Thus the process of phasor multiplication consists of the arithmetic multiplication of the magnitudes of the two phasors coupled with the addition of the two angles.

Division
Phasor division is defined by the equation

$$I_{\deg A_3} = (V_{\deg A_1})/(Z_{\deg A_2})$$

where $I = V/Z$ and $A_3 = A_1 - A_2$. The process of phasor division thus consists of the arithmetic division of the two phasor magnitudes coupled with the subtraction of the two angles, the order of subtraction being important.

4.2.2 Voltage–current relationships in phasor form
Voltage–current relationships in phasor form and the processes of phasor multiplication and division can be demonstrated by using the phasor relationships for inductive and capacitive reactance and susceptance. These are illustrated in figures 4.2 and 4.3.

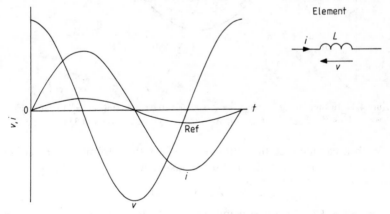

Figure 4.2 Instantaneous values: $i = 4 \sin(\omega t + 0)$, $v = 6 \sin(\omega t + \pi/2)$, ref $= \sin \omega t$. Phasor values: $I = 4_{\mathrm{rad}0}$, $V = 6_{\mathrm{rad}\pi/2}$, $X = V/I = 1.5_{\mathrm{rad}\pi/2}$, $B = I/V = 0.667_{\mathrm{rad}-\pi/2}$.

Where phasor analysis is understood, the whole of the phasor is frequently referred to by the appropriate upper case letter only as in figure 4.2; the presence of the angle is assumed and must of course be used in any calculation.

For an inductance the voltage drop leads the current by $\frac{1}{2}\pi$ radians.

The reactance phasor is therefore given by

$$X = X_{\mathrm{L\,rad\,\pi/2}} = V/I$$

where $X_\mathrm{L} = \omega L$, the inductive reactance as defined in §3.3.

The susceptance phasor is given by

$$B = B_{\mathrm{L\,rad}-\pi/2} = I/V$$

where $B_\mathrm{L} = 1/(\omega L)$, the inductive susceptance.

The current lags the voltage drop by $\frac{1}{2}\pi$ radians hence the phasor angle is $-\frac{1}{2}\pi$ radians; in other words the current phase relative to that of the voltage drop.

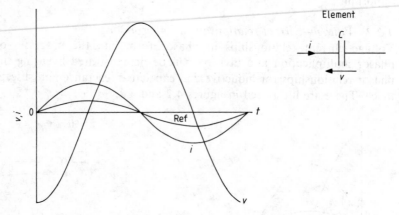

Figure 4.3 Instantaneous values: $i = 2\ \sin(\omega t + 0)$, $v = 6\ \sin(\omega t - \pi/2)$, ref $= \sin\ \omega t$. Phasor values: $I = 2_{\mathrm{rad}0}$, $V = 6_{\mathrm{rad}-\pi/2}$, $X = V/I = 3_{\mathrm{rad}-\pi/2}$, $B = I/V = 0.333_{\mathrm{rad}\pi/2}$.

With a capacitance the voltage drop lags the current by $\frac{1}{2}\pi$ radians, which means the reactance phasor is given by

$$X = X_{\mathrm{C\,rad}-\pi/2} = V/I$$

where $X_\mathrm{C} = 1/(\omega C)$. Since the current leads the voltage drop the susceptance phasor is given by

$$B = B_{\mathrm{C\,rad}\pi/2} = I/V$$

where $B_\mathrm{C} = \omega C$.

In phasor form the voltage–current relationships for the circuit elements and combinations of elements are all simple phasor ratios. The phase differences between voltage and current produced by the differential voltage–current relationships for inductance and capacitance are

accounted for by the phasor angles associated with reactance, susceptance, impedance and admittance phasors.

The phasor relationships for the circuit elements are summarized in
table 4.2.

Table 4.2 Phasor voltage–current relationships.

Element	Phasor	Phasor relationship	
Resistance	R	$R = V/I$	$R = 1/G$
Conductance	G	$G = I/V$	$G = 1/R$
Reactance	X	$X = V/I$	$X = 1/B$
Susceptance	B	$B = I/V$	$B = 1/X$
Impedance	Z	$Z = V/I$	$Z = 1/Y$
Admittance	Y	$Y = I/V$	$Y = 1/Z$

Program filename 'PH' is a series of frames illustrating phasor
properties for a number of different circuit element examples. The
program shows both the instantaneous waveforms and the related phasor
models to firmly establish the link between the two. The understanding
of this link is important because problems will ultimately need to be
interpreted in physical terms, the models being only an aid to analysis.
The interpretation of models which are used in analysing physical
systems is an essential part of the process.

4.2.3 Phasor addition

The process of phasor addition is derived from the result of adding two
sine waves of the same frequency, expressed by the equation

$$V_1 \sin(\omega t + A) + V_2 \sin(\omega t + B) = V_3 \sin(\omega t + C).$$

In phasor form this equation becomes

$$V_{1\,\mathrm{rad}A} + V_{2\,\mathrm{rad}B} = V_{3\,\mathrm{rad}C}.$$

Figure 4.4 illustrates the method of performing phasor addition which
is based on the following properties of sine waves.

For the three waves given above and shown in figure 4.4, when $t = 0$

$$V_1 \sin A + V_2 \sin B = V_3 \sin C$$

when $t = \frac{1}{4} P$, $\omega t = \frac{1}{2}\pi$

$$V_1 \cos A + V_2 \cos B = V_3 \cos C$$

since

$$V_1 \sin(A + \tfrac{1}{2}\pi) = V_1 \cos A$$

etc. Dividing these two equations gives

$$C = \tan^{-1}[(V_1 \sin A + V_2 \sin B)/(V_1 \cos A + V_2 \cos B)]$$

$$V_3 = (V_1 \sin A + V_2 \sin B)/\sin C.$$

The process of adding two phasors is therefore defined as producing a third phasor of magnitude and phase related to the two phasors added by the above equations.

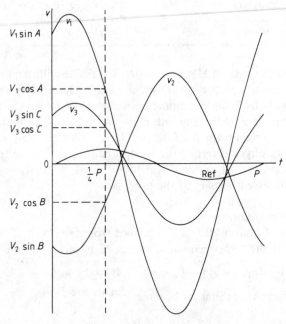

Figure 4.4

Subtraction

The relationship between the 'sign' and 'angle' of a phasor quantity is similar to that for a sine wave illustrated by the equation

$$V \sin(\omega t + A) = -[V \sin(\omega t + A \pm \pi \text{ radian})].$$

For a phasor the equivalent equation is

$$V \operatorname{rad} A = -[V \operatorname{rad}(A \pm \pi)]$$

or

$$V_{\deg B} = -(V_{\deg(B \pm 180)}).$$

For both the sine wave and the phasor, the negative sign is interpreted by the angle, the amplitude and magnitude, respectively, remaining the same.

Subtracting one phasor from another is the same as adding the 'negative' of the second phasor. As indicated above, this is equivalent to adding the second phasor with $180°$ added to its angle. This corresponds to the process of subtracting one sine wave from another by adding its antiphase wave.

4.2.4   Resolution of a phasor

The reverse process of addition is that of resolving the phasor into two component parts. The components which are of the greatest significance and use are the *reference* and *quadrature* components. In other words, those phasors with angles of 0 or 180 and 90 or -90 degrees respectively.

Applying the expressions in §4.2.3 to the addition of two phasors with angles of 0 and $90°$, respectively, is equivalent to adding the two resolved components.

For $A = 0°$ and $B = 90°$

$$\sin A = 0 \qquad \sin B = 1 \qquad \cos A = 1 \qquad \cos B = 0.$$

Therefore

$$V_2 = V_3 \sin C \qquad V_1 = V_3 \cos C.$$

Hence the reference and quadrature components have magnitudes of $V_3 \cos C$ and $V_3 \sin C$, respectively.

If $C > 90°$ or $< -90°$ the reference component has an angle of $180°$. If $C < 0$ the quadrature component has an angle of $-90°$.

Further properties of the two components are

(i) $V_3 = (V_1^2 + V_2^2)^{1/2}$.

(ii) $C = \tan^{-1}(V_2/V_1)$.

(iii) It can be seen that the process of phasor addition in §4.2.3 is the same as adding the reference and quadrature components separately and then combining them, as in (i) and (ii). This property of the reference and quadrature components is particularly useful for the addition of several phasors. The process is particularly straightforward because all the reference and all the quadrature components are respectively parallel and can therefore be added algebraically. In other words, those reference components with angles of $\pm 180°$ are subtracted, as are the quadrature components with angles of $-90°$. This technique is used frequently in the phasor programs.

In relation to sinusoidal sources the reference and quadrature components correspond to sine and cosine waves respectively and any sine wave can thus be represented by the sum of a sine and cosine wave of the appropriate magnitudes. The addition of component sine waves is illustrated by the waves in figure 3.7 and the corresponding phasor addition in figure 4.6. This property can be used in circuit analysis where there is advantage in considering the contribution of each component to the total response. An example of the application of this technique is the analysis in §4.6.1.

Program filename 'PHC' performs phasor calculations in accordance with the processes defined above. The program can be used to solve a wide range of problems which involve phasor manipulation and combination. Examples of the use of the program are given in §4.4.3 and §4.5.7.

4.2.5 Graphical representation of phasors

The relative magnitudes and angles of phasors can be illustrated by drawing them as vectors on a four-quadrant diagram, as shown in figure 4.5. The length of the vector from the axes origin represents the magnitude of the phasor and the angle of the vector relative to the reference axis is equal to the phasor angle. Positive angles are drawn anticlockwise from the reference axis and negative angles clockwise.

The process of phasor addition can also be performed graphically. This is shown by the broken line in figure 4.5 which represents the addition of phasor P_2 in magnitude and direction. The process is the same as vector addition and is the basis of vector models used in steady-state sine-wave analysis. The quantities which are modelled by vectors are not themselves vectors.

Graphical representation produces a simple method of addition and provides a visual representation which can assist perception. In analytical terms it is limited.

Program filename 'GRPH' performs the graphical addition of two phasors. The program can be used as an alternative to program filename 'AS' as a problem-solving tool for practice problem 3.2 and other similar problems.

Figure 4.6 shows the addition of a reference and quadrature phasor and is equivalent to the sine-wave addition shown in figure 3.7.

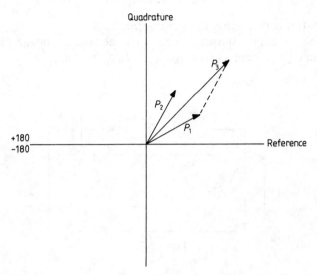

Figure 4.5 $P_1 = 100_{\deg 30}$, $P_2 = 100_{\deg 60}$, $P_3 = 193_{\deg 45}$, $P_3 = P_1 + P_2$.

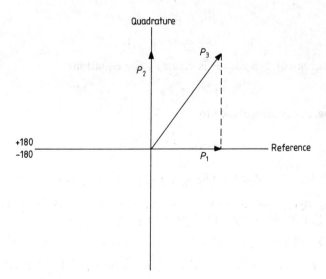

Figure 4.6 $P_1 = 3_{\deg 0}$, $P_2 = 4_{\deg 90}$, $P_3 = 5_{\deg 53.1}$, $P_3 = P_1 + P_2$.

⟩ **4.3   Impedance and admittance phasors**

4.3.1   Series combination of circuit elements

For the series circuit shown in figure 4.7 the source voltage phasor is $V_{\text{deg}A}$, and the circuit current phasor, $I_{\text{deg}0}$, is common to all three elements.

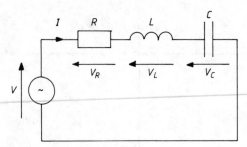

Figure 4.7

The three element phasor voltages are therefore given by

$$V_R = R_{\text{deg}0}I_{\text{deg}0}$$

$$V_L = X_{\text{Ldeg}90}I_{\text{deg}0}$$

$$V_C = X_{\text{Cdeg}-90}I_{\text{deg}0}.$$

The source voltage phasor is given by the equation

$$V_{\text{deg}A} = V_R + V_L + V_C$$

and the circuit impedance by

$$Z = V_{\text{deg}A}/I_{\text{deg}0}.$$

Hence

$$Z_{\text{sdeg}A} = (R_{\text{deg}0} + X_{\text{Ldeg}90} + X_{\text{Cdeg}-90}).$$

For this particular circuit the reference and quadrature components can be combined using the relationship in §4.2.4 to give

$$Z_s = [R^2 + (X_L - X_C)^2]^{1/2} = (R^2 + \{(\omega L) - [1/(\omega C)]\}^2)^{1/2}$$

$$A = \tan^{-1}[(X_L - X_C)/R] = \tan^{-1}(\{(\omega L) - [1/(\omega C)]\}/R).$$

It can be seen from the expression for the impedance magnitude that Z_s is a minimum and equal to the circuit resistance R when $X_L = X_C$. This is a resonant condition when for a given source voltage the source current will be a maximum. Resonance is covered in more detail in §4.9.

In general the impedance phasor for a series circuit is obtained by adding the resistance and reactance phasors. The sign of the phasor angle, A, depends on whether the inductive reactance is greater or less than the capacitive reactance.

4.3.2 Parallel combination of circuit elements

For the parallel circuit shown in figure 4.8 the circuit current phasor is $I_{\mathrm{deg}A}$ and the source voltage phasor $V_{\mathrm{deg}0}$ is common to all three elements.

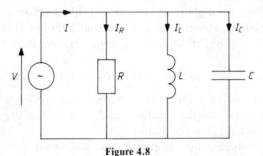

Figure 4.8

The three element phasor currents are therefore given by

$$I_R = G_{\mathrm{deg}0} V_{\mathrm{deg}0}$$

$$I_L = B_{L\,\mathrm{deg}-90} V_{\mathrm{deg}0}$$

$$I_C = B_{C\,\mathrm{deg}90} V_{\mathrm{deg}0}.$$

The source current phasor is given by the equation $I_{\mathrm{deg}A} = I_R + I_L + I_C$ and the circuit admittance by $Y = I_{\mathrm{deg}A}/V_{\mathrm{deg}0}$. Hence

$$Y_{\mathrm{p\,deg}A} = (G_{\mathrm{deg}0} + B_{L\,\mathrm{deg}-90} + B_{C\,\mathrm{deg}90}).$$

For this particular circuit the reference and quadrature components can be combined using the relationships in § 4.2.4 to give

$$Y_{\mathrm{p}} = [G^2 + (B_C - B_L)^2]^{1/2} = (G^2 + \{(\omega C) - [1/(\omega L)]\}^2)^{1/2}$$

$$A = \tan^{-1}[(B_C - B_L)/G] = \tan^{-1}(\{(\omega C) - [1/(\omega L)]\}/G).$$

When $B_C = B_L$, Y_{p} is a minimum and equal to G. This is a resonant condition when for a given source voltage the source current will be a minimum.

In general, the admittance phasor for a parallel circuit is produced by adding the conductance and susceptance phasors. The sign of the phasor angle, A, depends on whether the capacitive susceptance is greater or less than the inductive susceptance.

⟩ **4.4   Complex number representation**

As previously stated, the concentration in this treatment is on computer application to circuit theory and it seems likely that analytical tools will change as computer techniques spread more widely. Analytical techniques based on the algebra of complex variables are very powerful and so much a part of present circuit theory that although they are not used in the computer problem-solving programs they are introduced here for completeness.

Complex variable techniques fall into two broad areas where

(i)   voltages, currents and circuit elements are modelled for a specific frequency as complex numbers and combined in the same manner as the phasors described above;

(ii) circuit elements are modelled and combined and the resulting functions analysed in terms of a complex frequency variable.

The two techniques can be linked together in more comprehensive analytical treatments of circuit theory but the description which follows here is limited to the use of complex number algebra as an analytical alternative to computer manipulation of phasors.

4.4.1   Complex number algebra

The properties of complex numbers which are used in steady-state sine-wave analysis are virtually identical to those for the phasors described above. This section describes briefly the aspects of complex number algebra which are required for circuit analysis.

Complex numbers have real and imaginary parts. These can be expressed separately in rectangular form or combined as a polar expression. The rectangular form corresponds to phasor components and the polar form to the phasor magnitude and angle, the angles being equal when expressed in radians. Both forms of complex numbers can be represented graphically on a complex plane consisting of a horizontal axis for real components and a vertical axis for imaginary ones. An example is shown in figure 4.9 and the two forms are related by the equation

$$a + jb = M \exp(jA).$$

Relating the parameters in the two forms gives

$$a = M \cos A \qquad b = M \sin A$$

$$M = (a^2 + b^2)^{1/2} \qquad A = \tan^{-1}(b/a).$$

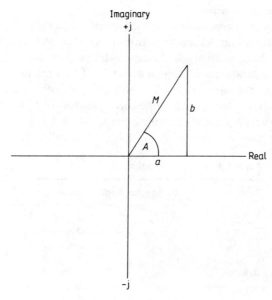

Figure 4.9

Addition
The process of complex number addition consists of adding the real and imaginary parts separately; hence the rectangular form is more convenient. The process of addition is shown by the equations below.

If $N_1 = a_1 + jb_1$ and $N_2 = a_2 + jb_2$ then

$$N_1 + N_2 = (a_1 + a_2) + j(b_1 + b_2).$$

Multiplication
The process of complex number multiplication is demonstrated by the polar form.

If $N_1 = M_1 \exp(j A_1)$ and $N_2 = M_2 \exp(j A_2)$ then

$$N_1 N_2 = M_1 M_2 \exp[j(A_1 + A_2)].$$

Division

Complex number division can be similarly illustrated by the equation

$$N_1/N_2 = (M_1/M_2)\exp[j(A_1 - A_2)].$$

4.4.2 Complex number models

The complex number models for the circuit elements are collected in table 4.3. The voltage–current relationships in complex number form are the same as those defined for the computer phasors in §4.2.2.

Inductive reactance is $j X_L$, with an angle of $\frac{1}{2}\pi$ radians in polar form, and capacitive reactance is $-j X_C$, with an angle of $-\frac{1}{2}\pi$ radians.

The various models are combined as described in §§4.3.1 and 4.3.2. Hence for the particular examples given $Z = R + j(X_L - X_C)$ for the series circuit and $Y = G + j(B_C - B_L)$ for the parallel circuit.

Table 4.3 Complex number models.

Property	Rectangular form	Polar form
Voltage	$V_R \pm j V_Q$	$V \exp(\pm jA)$
Current	$I_R \pm j I_Q$	$I \exp(\pm jA)$
Resistance	R	$R \exp(j0)$
Conductance	G	$G \exp(j0)$
Reactance	$\pm jX$	$X \exp(\pm \frac{1}{2}j\pi)$
Susceptance	$\mp jB$	$B \exp(\mp \frac{1}{2}j\pi)$
Impedance	$R \pm jX$	$Z \exp(\pm jA)$
Admittance	$G \mp jB$	$Y \exp(\mp jA)$

4.4.3 Comparison of computer and complex number methods

One aspect of the comparative qualities of computer and complex number methods can be illustrated by considering series and parallel equivalent circuits.

The problem can be stated by referring to the circuits shown in figure 4.10. Values of R_s, L_s, R_p and L_p are required which make the two circuits equivalent with respect to their terminals.

A specific example to demonstrate the two methods can be stated as requiring the values of L_s and R_s which will make the series circuit equivalent to a parallel connection of a resistance of 1000 Ω and an inductance of 159 mH, at a frequency of 1000 Hz.

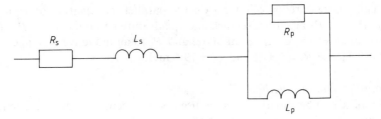

Figure 4.10

Computer method
Program filename 'REA' computes reactance and susceptance phasors for given element values at a specified frequency. The program additionally computes element values for given reactance or susceptance phasors. The printout with the relevant values for this example is shown in figure 4.11. The operation of program filename 'PHC' and the printout related to this example are given in figure 4.12.

SYMBOL	COMPONENT VALUE	FREQUENCY	REACTANCE	ANGLE	SUSCEPTANCE	ANGLE
L or C	henry or farad	hertz	ohm	degree	siemen	degree
L	0.159	1000	999	+90	1.001E-3	-90
L	7.958E-2	1000	500	+90	2E-3	-90

Figure 4.11

PHASOR CALCULATIONS

This program combines phasors as described by the processes in section 4.2.

The combination is performed on any two phasors from a defined set, the overall size of which is only limited by screen space.

The phasors are identified by number. Selecting a number greater than the currently displayed maximum produces a prompt for new input.

The phasor operations available are, addition(+), subtraction(-), multiplication(*), division(/) and resolution(C) into reference and quadrature components.

The operation is chosen by pressing the appropriate character key indicated in brackets above.

The phasor 1deg0 is included to facilitate reciprocal operations.

CALCULATION	PHASOR								
P3=P1*P2	No.	Magnitude	deg.	No.	Magnitude	deg.	No.	Magnitude	deg.
P4=P1+P2	P0	1	0.000						
P5=P3/P4	P1	1000	0.000						
P6=RCP5	P2	1000	90.00						
P7=QCP5	P3	1E6	90.00						
	P4	1414	45.00						
	P5	707.1	45.00						
	P6	500	0.000						
	P7	500	90.00						

Figure 4.12

The phasors are as follows: P_5 is the parallel combination of P_1 and P_2, P_6 and P_7 are the reference and quadrature components of P_5, P_6 corresponds to a resistance of 500 Ω and P_7 to an inductive reactance of 500 Ω. Hence $R_s = 500\ \Omega$ and $L_s = 79.6$ mH.

Complex number method

Complex number analysis can be used with specific values in a similar manner to the computer method above. A particular strength of algebraic methods is, however, in producing general relationships which can be used for a range of similar problems.

This is the method adopted for this example which produces the following analysis.

The two impedances are related by the equation

$$R_s + j\omega L_s = R_p j\omega L_p / (R_p + j\omega L_p).$$

The real and imaginary components are equated separately by

$$R_s = R_p \omega^2 L_p^2 / (R_p^2 + \omega^2 L_p^2)$$

$$L_s = L_p R_p^2 / (R_p^2 + \omega^2 L_p^2).$$

Hence for any particular set of element values and frequency the equivalents can be found. For this example the equations produce the values above.

Both methods are similarly applicable to circuits with resistance and capacitance.

⟩ 4.5 Phasor forms of analytical techniques and processes

Chapters 1 and 2 included a range of analytical techniques and processes which were limited to resistance-only circuits. These processes can now be extended to steady-state sine-wave analysis using phasor models which are governed by the same relationships that previously only applied to resistances. Since the electrical principles of the techniques and processes have already been covered it is sufficient in most cases to simply restate them in phasor form.

The overall strategy for steady-state sine-wave problem solving can be summarized by the block diagram in figure 4.13.

The parameters in the remaining sections of this chapter should be assumed to be phasors where appropriate unless otherwise stated.

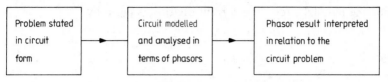

Figure 4.13

4.5.1   Impedances in series and parallel

The relationships for the series and parallel connection of impedances are given in figure 4.14. The circuit elements can be modelled as either impedances or admittances but the series combination is more appropriate in impedance form and the parallel combination is better suited to admittances. Each impedance and admittance is itself a combination of circuit elements but the rules of combination apply equally well to pure reactances or susceptances.

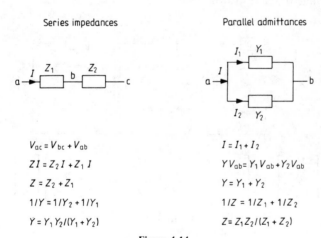

Figure 4.14

4.5.2   Voltage and current division

Voltage and current division can be expressed in phasor form by referring to the circuits in figure 4.14.

For the series circuit the voltage division is given by the two equations.

$$V_{bc} = [Z_2/(Z_1 + Z_2)] V_{ac}$$

$$V_{ab} = [Z_1/(Z_1 + Z_2)] V_{ac}.$$

For the parallel circuit the current division is given by

$$I_1 = [Y_1/(Y_1 + Y_2)]I$$
$$I_2 = [Y_2/(Y_1 + Y_2)]I.$$

4.5.3 Superposition
The principle of superposition as described in §2.2 is equally applicable to circuits modelled in phasor form.

The principle is used widely in considering separately components of single-frequency phasor voltages and currents and different harmonic frequency components of non-sinusoidal waveforms.

4.5.4 Mesh and node analysis
Figures 4.15 and 4.16 show how the matrices for mesh and node analysis are formed in terms of circuit impedance and admittance phasors and how the matrix relates to the circuit.

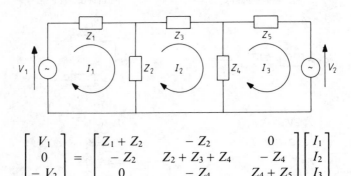

$$\begin{bmatrix} V_1 \\ 0 \\ -V_2 \end{bmatrix} = \begin{bmatrix} Z_1 + Z_2 & -Z_2 & 0 \\ -Z_2 & Z_2 + Z_3 + Z_4 & -Z_4 \\ 0 & -Z_4 & Z_4 + Z_5 \end{bmatrix} \begin{bmatrix} I_1 \\ I_2 \\ I_3 \end{bmatrix}$$

Figure 4.15

The processes are identical to those described for resistance-only circuits in §§2.3 and 2.4. The terms clockwise mesh currents and voltage sources acting clockwise are applicable to the phasor models in the same manner as previously described.

The impedance and admittance matrices for circuits with independent sources only are symmetrical about the diagonal. Using this fact can shorten the process of data input to problem-solving programs. Dependent sources upset this symmetry as shown below.

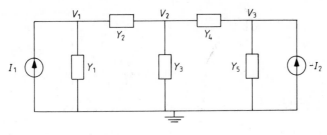

$$\begin{bmatrix} I_1 \\ 0 \\ -I_2 \end{bmatrix} = \begin{bmatrix} Y_1 + Y_2 & -Y_2 & 0 \\ -Y_2 & Y_2 + Y_3 + Y_4 & -Y_4 \\ 0 & -Y_4 & Y_4 + Y_5 \end{bmatrix} \begin{bmatrix} V_1 \\ V_2 \\ V_3 \end{bmatrix}$$

Figure 4.16

Dependent sources
The mesh circuit shown in figure 4.17 has an independent voltage source, V_1, and a dependent voltage source, ZI_3. This produces a modified impedance matrix with the voltage source represented by $-Z$, as shown.

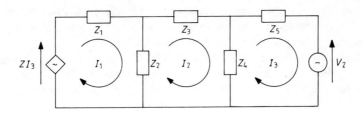

$$\begin{bmatrix} ZI_3 \\ 0 \\ -V_2 \end{bmatrix} = \begin{bmatrix} Z_1 + Z_2 & -Z_2 & 0 \\ -Z_2 & Z_2 + Z_3 + Z_4 & -Z_4 \\ 0 & -Z_4 & Z_4 + Z_5 \end{bmatrix} \begin{bmatrix} I_1 \\ I_2 \\ I_3 \end{bmatrix}$$

$$\begin{bmatrix} 0 \\ 0 \\ -V_2 \end{bmatrix} = \begin{bmatrix} Z_1 + Z_2 & -Z_2 & -Z \\ -Z_2 & Z_2 + Z_3 + Z_4 & -Z_4 \\ 0 & -Z_4 & Z_4 + Z_5 \end{bmatrix} \begin{bmatrix} I_1 \\ I_2 \\ I_3 \end{bmatrix}$$

Figure 4.17

The circuit shown in figure 4.18 has a dependent current source, YV_2, which likewise modifies the admittance matrix.

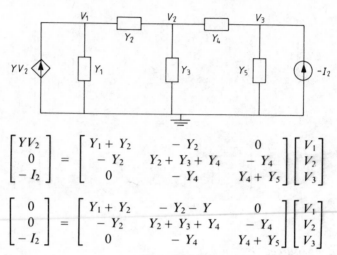

$$\begin{bmatrix} YV_2 \\ 0 \\ -I_2 \end{bmatrix} = \begin{bmatrix} Y_1 + Y_2 & -Y_2 & 0 \\ -Y_2 & Y_2 + Y_3 + Y_4 & -Y_4 \\ 0 & -Y_4 & Y_4 + Y_5 \end{bmatrix} \begin{bmatrix} V_1 \\ V_2 \\ V_3 \end{bmatrix}$$

$$\begin{bmatrix} 0 \\ 0 \\ -I_2 \end{bmatrix} = \begin{bmatrix} Y_1 + Y_2 & -Y_2 - Y & 0 \\ -Y_2 & Y_2 + Y_3 + Y_4 & -Y_4 \\ 0 & -Y_4 & Y_4 + Y_5 \end{bmatrix} \begin{bmatrix} V_1 \\ V_2 \\ V_3 \end{bmatrix}$$

Figure 4.18

Programs filenames 'MAZ' and 'NAY' are problem-solving mesh and node analysis programs for AC circuits. The numerical methods used are similar to those used for the resistance-only programs made possible by partitioning the matrices in terms of the reference and quadrature components which are determined separately and then recombined. This process is described briefly in Appendix A4.5.4.

Programs filenames 'MAZC' and 'NAYC' can be used as alternatives for circuits with independent sources only. These programs take in the data in component form and use the symmetry of the impedance and admittance matrices to reduce the input procedure.

4.5.5 Norton and Thevenin alternating current equivalents

Norton and Thevenin equivalents can be obtained for AC circuits in the same way as described in §2.6 for resistance-only circuits. The resulting parameter relationships are summarized in figure 4.19.

The programs required for computing Thevenin and Norton equivalents are under filenames 'MAZ' or 'NAY' and 'PHC'.

The procedure is typically as follows.

(i)   Determine I and I_{sc}, using program filename 'MAZ'.

(ii) Use program filename 'PHC' to determine Z and V_{oc} from the expressions $Z = Z_x/[(I_{sc}/I) - 1]$ and $V_{oc} = ZI_{sc}$.

     Alternatively, Z can be determined by using program 'PHC' to analyse the circuit with zero sources.

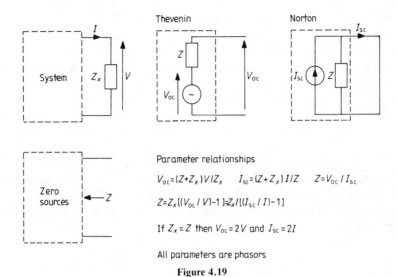

Parameter relationships

$$V_{oc} = (Z + Z_x)V/Z_x \qquad I_{sc} = (Z + Z_x)I/Z \qquad Z = V_{oc}/I_{sc}$$

$$Z = Z_x[(V_{oc}/V)-1] = Z_x/[(I_{sc}/I)-1]$$

If $Z_x = Z$ then $V_{oc} = 2V$ and $I_{sc} = 2I$

All parameters are phasors

Figure 4.19

Table 4.4 π and T equivalents (impedances).

π to T	T to π
$z_1 = Z_1 Z_2/(Z_1 + Z_2 + Z_3)$	$Z_1 = (z_1 z_2 + z_2 z_3 + z_3 z_1)/z_2$
$z_2 = Z_2 Z_3/(Z_1 + Z_2 + Z_3)$	$Z_2 = (z_1 z_2 + z_2 z_3 + z_3 z_1)/z_3$
$z_3 = Z_3 Z_1/(Z_1 + Z_2 + Z_3)$	$Z_3 = (z_1 z_2 + z_2 z_3 + z_3 z_1)/z_1$

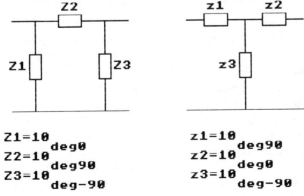

```
Z1=10
      deg0
Z2=10
      deg90
Z3=10
      deg-90
```

```
z1=10
      deg90
z2=10
      deg0
z3=10
      deg-90
```

Figure 4.20

4.5.6   π and T equivalence (impedances)

The relationships given for π and T circuit equivalence given in §2.10 can be restated in phasor form for impedance networks as shown in table 4.4.

As stated in §2.10, one of the applications of π and T equivalence is the analysis of three-phase circuits where the π and T circuits are drawn in delta and star form as described in §4.7.

Program filename 'PTZ' computes π and T circuit equivalents in phasor form. A printout for specific circuits is shown in figure 4.20.

4.5.7   An example in the use of problem-solving programs

This section illustrates the application of four of the problem-solving programs associated with this chapter. The problem solved is the computation of the values of the currents I_1 and I_2 in the circuit shown in figure 4.21.

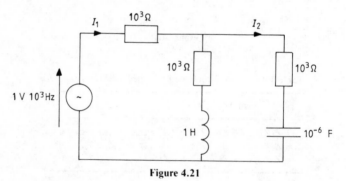

Figure 4.21

Three methods of solving the problem are described below.

(i)   The first method involves calculating the total impedance of the circuit. This produces the source current I_1, from which I_2 can be determined by applying the principle of current division to the two parallel branches of the circuit.

Program filename 'PHC' is used for the analysis and the printout is shown in figure 4.22.

The phasors are as follows: P_2 is the reactance of the inductance ('REA'), P_3 is the impedance of the inductive branch, P_4 is the reactance of the capacitance ('REA'), P_5 is the impedance of the capacitive branch, P_8 is the combined impedance of the parallel branches, P_9 is the total impedance of the circuit, $P_{10} = I_1$ and $P_{12} = I_2$.

CALCULATION			PHASOR								
			No.	Magnitude	deg.	No.	Magnitude	deg.	No.	Magnitude	deg.
P3=P1+P2			P0	1	0.000						
P5=P4+P1			P1	1000	0.000						
P6=P3*P5			P2	6283	90.00						
P7=P3+P5			P3	6362	88.96						
P8=P6/P7			P4	159.2	-90.00						
P9=P1+P8			P5	1013	-9.043						
P10=P8/P9			P6	6.442E6	71.91						
P11=P10*P3			P7	6442	71.91						
P12=P11/P7			P8	1000	-0.000						
P13=P3+P1			P9	2000	-0.000						
			P10	5E-4	0.000						
			P11	3.181	88.96						
			P12	4.938E-4	9.043						
			P13	6594	72.34						

Figure 4.22

(ii) An alternative method of solution is mesh analysis. The branch currents I_1 and I_2 correspond to the clockwise mesh currents for the circuit. Hence either of the mesh analysis program filenames 'MAZ' or 'MAZC' produce the results directly.

Program filename 'MAZ' requires the input in phasor form and some initial computation is required to process the data. The matrix for the circuit is shown in figure 4.23 together with the printout of the results.

The matrix impedances are equal to the phasors in figure 4.22 as follows: $Z_{11} = P_{13}$, $Z_{12} = -P_3$, $Z_{21} = -P_3$ and $Z_{22} = P_7$.

$$\begin{bmatrix} V_1 \\ V_2 \end{bmatrix} = \begin{bmatrix} Z_{11} & Z_{12} \\ Z_{21} & Z_{22} \end{bmatrix} \begin{bmatrix} I_1 \\ I_2 \end{bmatrix}$$

```
UNKNOWN CURRENTS

I(1)=5E-4deg0
I(2)=4.93E-4deg9.04
```

Figure 4.23

Program filename 'MAZC' receives data in component form and can be used on the circuit without calculating the matrix impedances. The circuit has the special property that its impedance is purely resistive and constant at all frequencies; this is demonstrated by the frequency response of the circuit in figure 4.43.

Program filename 'MAZC' has the facility for the analysis frequency to be changed and the mesh currents recalculated. Figures 4.24 and 4.25 show printouts from 'MAZC' at frequencies of 1000 and 100 Hz, respectively, showing that the current I_1 does not change with frequency.

```
CLOCKWISE MESH CURRENTS

I(1)=5E-4deg0
I(2)=4.94E-4deg9.04
```

Figure 4.24

```
CLOCKWISE MESH CURRENTS

I(1)=5E-4deg0
I(2)=2.66E-4deg57.9
```

Figure 4.25

(iii) The third method demonstrated for solving the problem involves modifying the circuit to one with a current source, using the relationships given in figure 4.19.

The modified circuit is shown in figure 4.26 and the problem can now be solved using the node analysis program filename 'NAYC'.

Two printouts from 'NAYC' are shown in figures 4.27 and 4.28 for frequencies of 1000 and 100 Hz respectively. In both results the voltage V_1 is 0.5_{deg0} V, indicating that the parallel $R-L$ and $R-C$ combination carries half of the source current and behaves as a pure resistance of 1000 Ω.

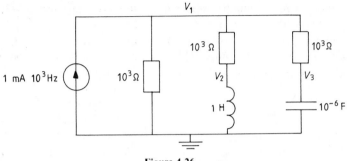

Figure 4.26

```
NODE VOLTAGES

V(1)=0.5deg0
V(2)=0.494deg9.04
V(3)=7.86E-2deg-81
```

Figure 4.27

```
NODE VOLTAGES

V(1)=0.5deg0
V(2)=0.266deg57.9
V(3)=0.423deg-32.1
```

Figure 4.28

⟩ 4.6 Further aspects of power in alternating current circuits

4.6.1 Power factor, analytical derivation
The computer-derived curve of power factor in terms of the phase difference between voltage drop and current is shown in figure 3.12. It can be demonstrated that the relationship is cosinusoidal by reference to the equivalent circuits shown in figure 4.29.

The circuit represents a sinusoidal voltage source supplying power to a load impedance, Z. The circuit current is taken as the phase reference and the instantaneous values of current and voltage are given by the equations

$$i = I_m \sin \omega t \qquad v = V_m \sin(\omega t + A).$$

The voltage source can be replaced by its reference and quadrature components as described in §4.2.4.

In phasor form the voltage sources are related by the equation $V = V_1 + V_2$ or more completely $V_{\text{mrad}A} = V_{\text{1mrad0}} + V_{\text{2mrad}\pi/2}$ where $V_{1m} = V_m \cos A$ and $V_{2m} = V_m \sin A$.

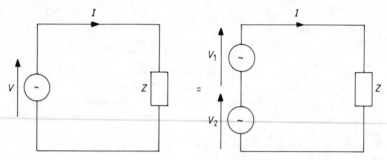

Figure 4.29

These equations correspond to the instantaneous voltage equations

$$V_m \sin(\omega t + A) = V_{1m} \sin \omega t + V_{2m} \sin(\omega t + \tfrac{1}{2}\pi)$$
$$= V_{1m} \sin \omega t + V_{2m} \cos \omega t.$$

Since the component sources are in series the current through each of them is the same and the instantaneous power expression is given by

$$p = V_{1m}I_m \sin^2 \omega t + V_{2m}I_m \sin \omega t \cos \omega t.$$

Using the relationships given in Appendix A6.4 for the average values of sine and cosine products, the expression for average power, P_{av}, is

$$P_{av} = \tfrac{1}{2}V_{1m}I_m = \tfrac{1}{2}V_m I_m \cos A = V_{rms}I_{rms} \cos A \text{ W}.$$

Hence the power factor is equal to $\cos A$. The sign of the phase angle, A, does not affect the power factor or the above analysis.

A more general analysis is given in Appendix A4.6.

4.6.2 Power transfer

The power transfer from a generator consisting of an alternating voltage source and an internal impedance can be considered by reference to the circuit shown in figure 4.30.

The generator internal impedance, Z_s, and load impedance, Z_x, can be considered to consist of series resistance and reactance and are given by the phasor equations

$$Z_s = R_s + X_{sdeg \pm 90}$$

$$Z_x = R_x + X_{xdeg \pm 90}.$$

The analysis for maximum power transfer from generator to load in this circuit is complicated by the fact that R_x and X_x can vary independently.

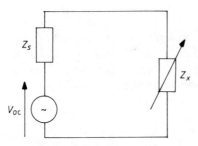

Figure 4.30

Since no average power is absorbed by reactive elements the maximum power is transferred to the load impedance when the power in the resistive component is a maximum. For a given value of resistance this will be a maximum when the current is a maximum.

In §4.3.1 it was shown that for a series circuit consisting of resistance and inductive and capacitive reactance the current is a maximum when the two reactances are equal. Hence for a given resistance the power transferred is a maximum when the two reactances are equal and of different types.

Under these conditions the effect of the series reactances cancel and the analysis in § 2.7 applies. This shows that maximum power transfer operates at 50% efficiency with equal parts of the source power being dissipated in the generator internal resistance and the load resistance.

Maximum power transfer for the circuit shown in figure 4.30 occurs therefore when the series resistive parts of Z_s and Z_x are equal and the series reactances are equal and of opposite types.

4.6.3 Power factor improvement

Power factor improvement, commonly called correction, is an aspect of power transfer which becomes relevant at relatively high power levels. The principles involved serve to demonstrate the application of some of the AC circuit theory given in this chapter.

Supply systems usually operate under constant amplitude conditions. The mains supply, for example, has a specified voltage. For a given average power the current will therefore be a minimum when the power factor is unity. From the point of view of losses and supply equipment capacity it is therefore desirable that systems operate at a minimum power factor and in many instances there is economical advantage in improving the power factor of individual items of equipment or complete commercial premises.

The principle of improving the power factor of a single-phase inductive load is shown in figure 4.31. The capacitance in parallel with the inductive load takes a leading quadrature current which partly or completely cancels the lagging quadrature current component of the inductive load as shown by the phasor diagram.

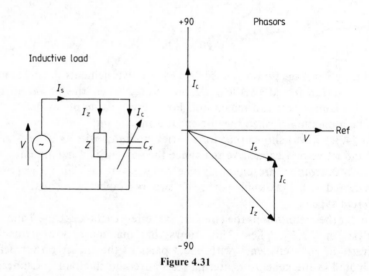

Figure 4.31

The graphs in figure 4.32 show how the supply current and power factor vary with capacitance.

Program filename 'PFI' illustrates the principle of power factor improvement. It uses the circuit phasor diagram to show the effect of connecting increasing values of capacitance in parallel with an inductive load. Program filename 'PF' produces the graphs and data shown in figure 4.32, for specific single-phase inductive loads, in response to user-defined parameters. Inductive loads are chosen as being more typical of the practical situation, although in principle correcting the power factor of a leading current load with an inductance would be equally valid.

Both aspects of power transfer considered in §§4.6.2 and 4.6.3 produce resonant conditions at maximum power transfer and unity power factor respectively. Resonance is discussed further in §4.9.

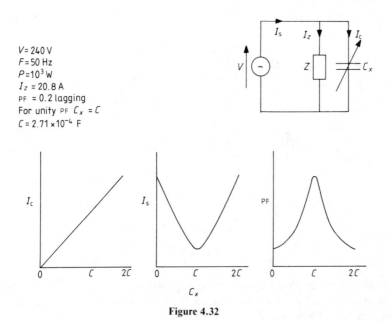

$V = 240$ V
$F = 50$ Hz
$P = 10^3$ W
$I_z = 20.8$ A
PF $= 0.2$ lagging
For unity PF $C_x = C$
$C = 2.71 \times 10^{-4}$ F

Figure 4.32

⟩ 4.7 Three-phase theory

It is not appropriate in this context to give a detailed explanation of the economic and technical advantages of three-phase systems. Three-phase supplies are used for large-scale energy distribution systems and the principles generally apply to relatively high power equipment. These subjects are consequently more the province of distribution and power system technology. This section is therefore limited to a basic description of three-phase supplies and circuits.

4.7.1 Three-phase supply circuits
A three-phase supply consists of three separate, sinusoidal, phase-related sources of the same amplitude and frequency. Each source is $120°$ out of

phase with the other two and the instantaneous voltages are given by the equations

$$v_1 = V_m \sin \omega t \qquad v_2 = V_m \sin(\omega t - \tfrac{2}{3}\pi) \qquad v_3 = V_m \sin(\omega t + \tfrac{2}{3}\pi).$$

Hence the phasors for the three sources are $V_{\text{rad}0}$, $V_{\text{rad}-2\pi/3}$ and $V_{\text{rad}2\pi/3}$, respectively.

The phase relationships for a three-phase supply are shown in figure 4.33.

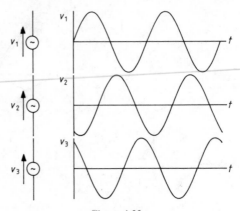

Figure 4.33

Instead of six conductors being required to connect the supply to a given load, the number is reduced by using star or delta methods of connection. These are shown in figure 4.34.

Star connection produces a three- or four-conductor supply depending on whether the supply neutral, which is the common connection between the three voltage sources, is connected to the load.

The delta supply is produced by connecting the three sources in series to form a closed circuit. This is permissible because the phase relationships produce a resultant voltage around the closed circuit of zero. The supply has three lines which are connected to the nodes at the junction of any two sources.

Phase and line voltages are the two voltage parameters used for three-phase supplies. The phase voltage is the voltage between any line and the neutral for a star supply. The line voltage is the voltage between any pair of lines. The line voltages for a star supply are therefore the difference between the two appropriate phase voltages. The relationship between the phase and line voltages is shown in the phasor diagram in figure 4.34. The phasor addition shows that the line voltages lead the phase voltages by 30° and are $\sqrt{3}$ times larger in magnitude. For a delta supply the line and phase voltages are the same. Program filename '3P' illustrates the above relationships. It demonstrates the three sinusoidal supplies and then relates the star and delta connections to the phasor diagrams. Programs with filenames 'AS', 'GRPH' and 'PHC' should be used to verify the magnitude and phase relationships between the phase and line values of voltage and current.

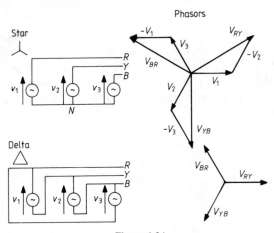

Figure 4.34

4.7.2 Three-phase load circuits

A three-phase load has three impedances connected in star or delta. Either type of load connection can be connected to either a star or delta supply. If the impedances are equal the load is said to be balanced. Figure 4.35 shows examples of three- and four-wire star and delta loads.

The solution of three-phase load currents presents no particular problems. Mesh analysis where necessary can be employed in a similar manner to its application in circuits where the sources are not phase related. Generally, more simple methods are appropriate.

For delta loads the voltage across each branch is known and the line currents are produced by adding the appropriate phase currents. For star loads the line and phase currents are equal. The voltage across the load phases depends on whether the neutral conductor is connected or whether the load is balanced.

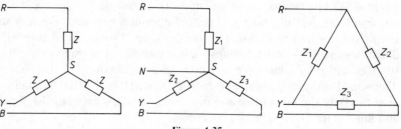

Figure 4.35

Delta-star equivalents

The relationships for π to T conversion can be applied to delta and star circuits. The logical labelling for the π and T circuits is different from that for the delta and star circuits (see figures 4.20 and 4.26). Program filename 'PTZ' is provided with a delta-star option and figure 4.36 shows a program printout.

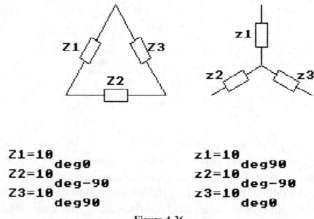

```
Z1=10                    z1=10
      deg0                     deg90
Z2=10                    z2=10
      deg-90                   deg-90
Z3=10                    z3=10
      deg90                    deg0
```

Figure 4.36

⟩ 4.8 Frequency responses

4.8.1 Introduction

So far in this chapter the circuit analysis has been in terms of a single frequency. It is important in a number of applications to consider the response of a system over a range of frequencies. Two examples with opposite requirements are filter circuits, which are designed to respond to a range of frequencies in a selective manner, whereas circuits designed to transfer signals from one part of an electronic system to another may be required to treat all frequencies as equally as possible. In both instances the frequency response of the circuits is an important indicator of their behaviour.

Frequency responses can relate a variety of parameters to frequency; in circuit theory terms, impedance, admittance and voltage ratio phasors are typical examples.

Figure 4.37 shows the frequency response of the impedance for a series $R-L-C$ circuit and figure 4.38 the response of the voltage ratio V_2/V_1 for the $C-R$ circuit. The graphs give both magnitude (M) and phase (A) information and the application will determine the relative significance of the two parameters.

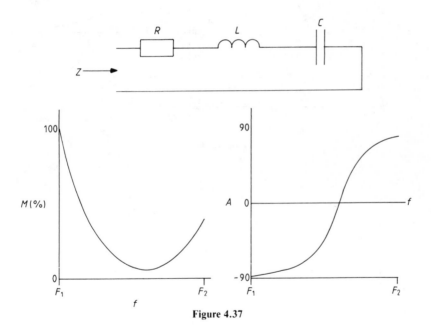

Figure 4.37

4.8.2 Logarithmic and linear scales

Where the response is required over a wide range of frequencies there is advantage in using logarithmic scales. The increments along the frequency axis for a logarithmic scale are proportional to the frequency value and therefore not constant, as is the case with linear scales.

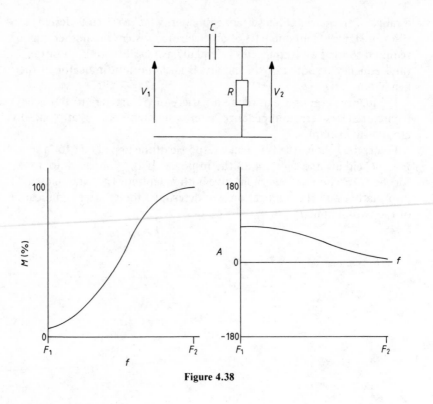

Figure 4.38

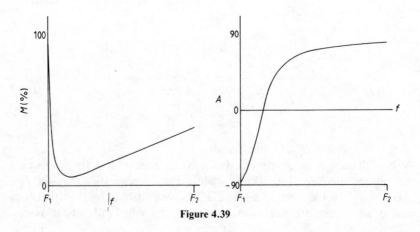

Figure 4.39

Logarithmic scales produce progressively smaller axis increments for equal frequency increments, the scale being proportional to the logarithm of the frequency.
Logarithmic scales necessarily change the shape of a characteristic. Figure 4.39 shows the impedance response for the $R-L-C$ series circuit over the same frequency range as shown in figure 4.37 but with a linear frequency scale.

4.8.3 Frequency response program example
Program filename 'FRP' computes frequency responses and then draws the appropriate graphs. Figure 4.40 shows the computer printouts for the response shown in figure 4.37.

FREQUENCY RESPONSE OF PHASOR FUNCTIONS

This program plots the freqency response of phasor functions, giving graphs of both magnitude and angle against frequency.

The functions may be impedances, admittances or unspecified phasors.

The first part of the process is to combine the phasors to produce the required function.

CALCULATION		R,X or Z PHASORS AT 1Hz							
P4=P1+P2		No.	Magnitude	deg.	No.	Magnitude	deg.	No.	Magnitude	deg.
P5=P3+P4		P0	1	0.000						
		P1	1000	0.000						
		P2	6.203	90.00						
		P3	1.592E5	-90.00						
		P4	1000	0.360						
		P5	1.592E5	-89.64						

FREQUENCY	PHASOR		FREQUENCY	PHASOR		FREQUENCY	PHASOR	
	Magnitude	deg.		Magnitude	deg.		Magnitude	deg.
10	1.588E4	-86.39	47.86	3186	-71.70	229.1	1247	36.67
10.96	1.448E4	-86.04	52.48	2882	-69.70	251.2	1376	43.37
12.02	1.32E4	-85.66	57.54	2604	-67.42	275.4	1526	49.06
13.18	1.203E4	-85.23	63.1	2349	-64.81	302	1697	53.80
14.45	1.097E4	-84.77	69.18	2117	-61.81	331.1	1887	57.99
15.85	9993	-84.26	75.86	1905	-58.34	363.1	2097	61.52
17.38	9104	-83.69	83.18	1713	-54.28	398.1	2327	64.55
19.05	8293	-83.07	91.2	1541	-49.53	436.5	2580	67.19
20.89	7553	-82.39	100	1388	-43.93	478.6	2856	69.50
22.91	6877	-81.64	109.6	1258	-37.33	524.8	3157	71.53
25.12	6259	-80.81	120.2	1150	-29.61	575.4	3486	73.33
27.54	5694	-79.89	131.8	1069	-20.76	631	3845	74.92
30.2	5178	-78.86	144.5	1018	-10.92	691.8	4237	76.35
33.11	4706	-77.73	158.5	1000	-0.480	758.6	4665	77.62
36.31	4274	-76.47	173.8	1015	9.985	831.8	5133	78.77
39.81	3879	-75.06	190.5	1063	19.90	912	5645	79.80
43.65	3517	-73.48	208.9	1142	28.85	1000	6205	80.73

Figure 4.40

⟩ **4.9   Resonance**

4.9.1   Resonance for particular circuits
The inductance and capacitance voltages for the series circuit and currents for the parallel circuit shown in §3.4 are in antiphase. This means that a current maximum for the inductance coincides with a capacitance voltage zero and a voltage maximum for the capacitance with an inductance current zero. In stored energy terms this means that either element has its maximum stored energy when the energy stored in the other one is zero. The maximum values of stored energy are equal when

$$\tfrac{1}{2} L I_m^2 = \tfrac{1}{2} C V_m^2$$

but

$$V_m = I_m/(\omega C)$$

therefore the maximum values of stored energy are equal when

$$\omega = (LC)^{-1/2} \qquad F_r = 1/[2\pi(LC)^{1/2}] \text{ Hz.}$$

F_r is the resonant frequency of the circuit and corresponds to the condition when the stored energy oscillates between the inductance and capacitance, with the total stored energy at any instant being constant and shared between the two elements.

At the resonant frequency the inductive and capacitive reactances are equal and the antiphase voltages and currents in the two circuits

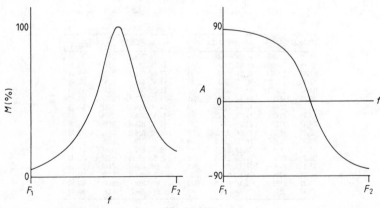

Figure 4.41

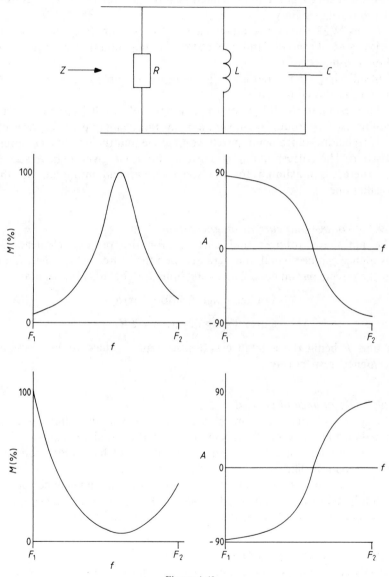

Figure 4.42

therefore cancel. This makes them appear to the circuit sources to consist of the resistance only.

Figure 4.37 shows the impedance variation with frequency for the series $R-L-C$ circuit and the corresponding admittance response is shown in figure 4.41.

Similar impedance and admittance responses for the parallel $R-L-C$ circuit are shown in figure 4.42.

The resonant condition for these series and parallel circuits can be summarized as producing maximum admittance and impedance, respectively, the admittance being purely conductive and the impedance purely resistive. The corresponding circuit currents for a given voltage source are therefore maximum for the series circuit and minimum for the parallel one.

4.9.2　Voltage and current magnification

The antiphase voltages and currents for the reactive elements at resonance can be several times greater than the source values, depending on the circuit parameters. The expressions for the magnification are

$$\text{Voltage magnification} = X/R$$

$$\text{Current magnification} = B/G$$

X and B being the reactance and susceptance values at the resonant frequency, respectively.

4.9.3　Resonance in general

The circuits considered in §4.9.1 have maximum admittance or impedance at the same frequency at which the circuits behave as pure resistances. The circuits in figures 4.43 and 4.44 have more complex characteristics. The circuit in figure 4.43 has an impedance which is purely resistive and constant at all frequencies. The impedance for the circuit in figure 4.44 has marked frequency dependence but the maximum value does not exactly coincide with zero phase difference.

4.9.4　Selectivity and Q factor

Selectivity is an important quality in applications where circuits are used for frequency discrimination. A measure of the frequency-selecting qualities of a circuit is its Q factor. There are a number of different ways of expressing the Q factor of a circuit. In relation to selectivity it is given

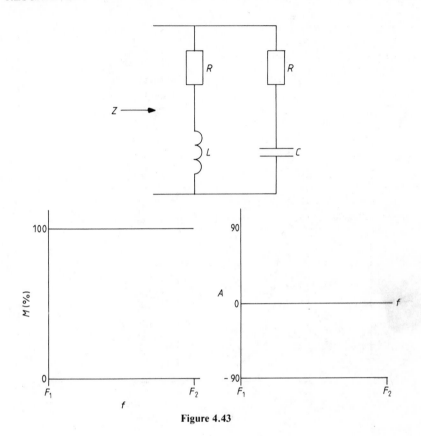

Figure 4.43

by the ratio

frequency for maximum response/bandwidth.

Bandwidth is the difference in the frequencies at which the response is $1/\sqrt{2}$ of its peak value.

Using these relationships the circuits in figures 4.43 and 4.44 have Q factors of 0 and approximately 2, respectively.

Q factor in terms of voltage magnification and energy
Alternative expressions for Q factor can be illustrated in relation to the series resonant circuit given in figure 4.37. Equating the circuit Q factor to the voltage magnification at resonance gives the expressions

$$Q = \omega_r L/R = 1/(\omega_r CR)$$

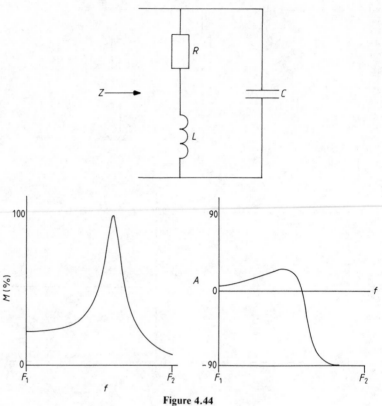

Figure 4.44

where ω_r is the resonant value given by

$$\omega_r = (LC)^{-1/2}.$$

Combining these relationships gives

$$Q = (L/C)^{1/2}/R.$$

Q factor can be related to energy by the ratio at resonance

maximum stored energy in either reactive element/the energy dissipated per cycle by the resistance.

If I_m is the maximum current, then the energy ratio using the inductance is

$$\tfrac{1}{2}L(I_m)^2/(2I_{rms}^2 R\pi/\omega_r) = \omega_r L/2\pi R = Q/2\pi.$$

> **4.10   Practice problems**

1   Perform the following phasor calculations using program filename 'PHC'.

   (i) $10_{deg0} + 30_{deg90} + 40_{deg-90}$

   (ii) $[(10_{deg0} + 30_{deg90})(40_{deg-90})](10_{deg0} + 30_{deg90} + 40_{deg-90})^{-1}$

   (iii) $(\{[(10_{deg0})(30_{deg90})]/(10_{deg0} + 30_{deg90})\}(40_{deg-90}))$
          $(\{[(10_{deg0})(30_{deg90})]/(10_{deg0} + 30_{deg90})\} + 40_{deg-90})^{-1}$

   (iv) $\{[(10_{deg0})(30_{deg90})] + [(30_{deg90})(40_{deg-90})] +$
          $[(40_{deg-90})(10_{deg0})]\}/10_{deg0}.$

2   Use as appropriate programs with filenames 'PHC', 'GRPH' and 'REA' to repeat the practice problems in §3.10.

3   Determine the impedance phasors for the four circuits shown in figure 4.45, at frequencies of 10, 100 and 1000 Hz. Consider the ease with which the four circuit models could be realized in practice.

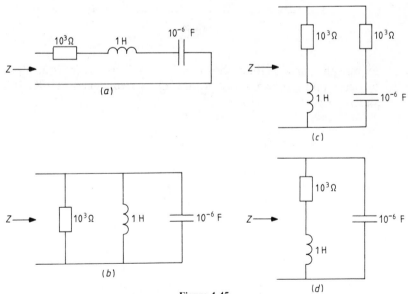

Figure 4.45

4   (i)   For the bridge circuit in figure 4.46, show that $V_{ab} = 0$ when $Z_x = Z_1 Z_2 / Z_3.$

(ii)   In a particular circuit, $V_{ab} = 0$ with $Z_1 = Z_2 = 1000_{deg0}$ and $Z_3 = 1000_{deg-45}$. Determine Z_x.

(iii) If Z_x consists of series resistance (R_s) and inductance (L_s), determine R_s and L_s at 50 Hz.

(iv)   Repeat (iii) to determine the parallel equivalent values, R_p and L_p.

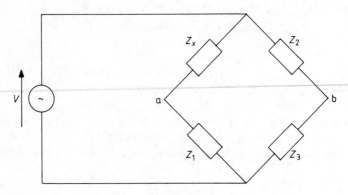

Figure 4.46

5   Express the parallel circuit shown in figure 4.47 as a series equivalent by determining the values of R_s and C_s at frequencies of 100, 10^4 and 10^6 Hz.

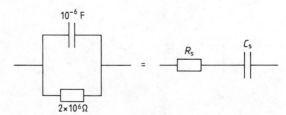

Figure 4.47

6   (i)   For the bridged T circuit in figure 4.48, show that $V_{ab} = 0$ when $Z_x + Z_1^2/Z_2 + 2Z_1 = 0$.

       (ii) In a particular circuit, $V_{ab} = 0$ with $Z_1 = 100_{deg-90}$ and $Z_2 = 100_{deg0}$. Determine Z_x.

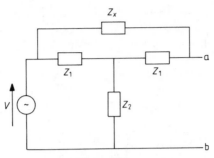

Figure 4.48

7 (i) Determine the four mesh currents for the circuit shown in figure 4.49, at voltage source frequencies of 25, 50 and 75 Hz. The phase of the voltage source is the reference phase.

(ii) At each frequency obtain the current in the 1000 Ω resistance and the source current.

(iii) Perform a node analysis on the circuit to obtain the voltage across the 1000 Ω resistance at frequencies of 25, 50 and 75 Hz and hence confirm the results of (ii).

Explain the circuit behaviour.

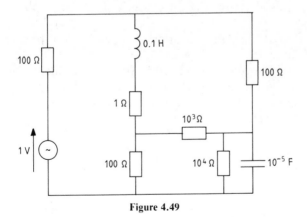

Figure 4.49

8 Compute the four node voltages for the bridged T network shown in figure 4.50 at a frequency of 10^6 Hz. The phase of the current source is the reference phase. Use the computed value of V_1 to replace the current source with a voltage source which will produce the same

circuit conditions, then perform a mesh analysis on the circuit. Compare the node and mesh analyses at frequencies near 10^6 Hz with the condition $V_{ab} = 0$ in problem 6.

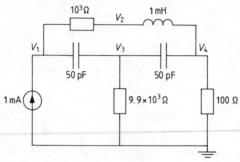

Figure 4.50

9   (i)   For the bridged T circuit (whole circuit—100 Ω resistance) in figure 4.50, determine the equivalent π circuit at a frequency of 10^5 Hz.

     (ii)   Determine the Thevenin and Norton equivalents for the bridged T circuit at a frequency of 10^5 Hz.

10   (i)   Derive the matrix equation for V_1 and V_2 for the circuit shown in figure 4.51.

     (ii)   Use the equation derived in (i) to determine V_i and V_o and hence the gain (V_o/V_i) for the amplifier equivalent circuit shown in figure 4.52, at frequencies of 10 kHz, 1 MHz and 10 MHz. $g_m = 5$ mA V^{-1}.

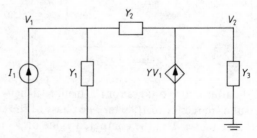

Figure 4.51

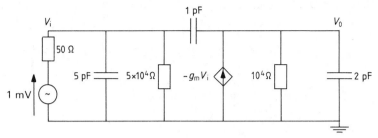

Figure 4.52

11 Calculate the phase and line currents for the three-phase delta load shown in figure 4.53. The supply has a line voltage of 440 V at 50 Hz, with a phase sequence RYB. The RY line voltage is the phase reference. Determine the average power in each phase and the total average power of the load.

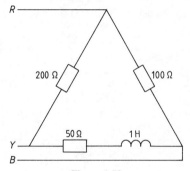

Figure 4.53

12 Use delta–star conversion on the circuit shown in figure 4.53 to determine the equivalent star load. Using mesh analysis on the equivalent star load, determine the three line currents and compare them with the answer to problem 11.

13 A 440 V line, 50 Hz three-phase supply, with phase sequence RYB, supplies the unbalanced three-wire load shown in figure 4.54. Use (i) mesh analysis and (ii) star–delta conversion to determine the current and average power in each phase. The RY line voltage is the phase reference. The impedance phasors are $Z_1 = 10_{deg0}$, $Z_2 = 20_{deg45}$ and $Z_3 = 10_{deg-30}$. (iii) Determine the voltage between the load star point and the supply neutral.

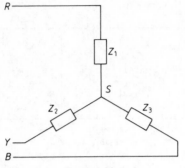

Figure 4.54

14 Determine the currents I_1 and I_2 in the coupled inductance circuit shown in figure 4.55 and the phase of the voltage drop across the 100 Ω resistance relative to the phase of the source voltage. The mutual inductance is 0.4 H. Calculate the power supplied by the source and that absorbed by the 100 Ω resistance.

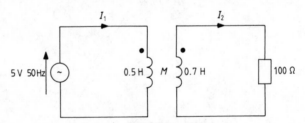

Figure 4.55

15 Obtain frequency response curves for the circuits shown in figure 4.45, showing the variation of impedance and admittance over the frequency range 10–1000 Hz for both logarithmic and linear frequency scales.

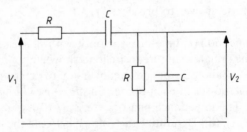

Figure 4.56

16 Display the voltage transfer ratio, V_2/V_1, for the circuit shown in figure 4.56, over the frequency range 10–1000 Hz. $R = 1000\ \Omega$, $C = 10^{-6}$ F.

⟩ Chapter 5

⟩ Circuit Analysis and Responses in the Time Domain

⟩ 5.1 Initial, transient and steady-state responses

General properties

The subject of initial and transient responses which occur when the state of a system is suddenly changed was introduced in §3.5 and is considered in more detail in this chapter.

The separation of transient and steady-state responses by their traditional association with time- and frequency-domain analysis, respectively, may mask the fact that systems and their circuit models have just one continuous response. Responses of stable systems change smoothly, albeit with different rates, from the initial response at the instant the system is changed, to the steady-state response when the change is imperceptible and the response is regarded as a constant or a repetitive function of time.

The computer methods used in this chapter produce initial, transient and steady-state time-domain circuit responses and by displaying the response at different times the continuous transition from initial to steady-state response can be demonstrated.

Initial response: convention and initial values

The analysis convention is adopted of using $t = 0_-$ for an instant in time immediately prior to $t = 0$ and $t = 0_+$ for a time immediately after zero. This differentiates between instances in time either side of zero while the response time remains at $t = 0$. The initial response is thus the response at the instant $t = 0_+$.

The initial values of circuit parameters of relatively simple circuits can

144

usually be determined by inspection using the circuit element voltage–current relationships. In more complex circuits initial responses can be determined by replacing inductances carrying current I_0 at $t = 0_-$ by current sources of I_0 and capacitances with voltage V_0 at $t = 0_-$ by voltage sources of V_0, both being zero sources or otherwise as appropriate. The circuit can then be analysed for the initial response at $t = 0_+$, as a circuit with sources and resistance only, by a suitable method. The equivalent circuit is only valid instantaneously.

Transient response: transient effect and transient
In this computer treatment of time-domain analysis the transient response of a system has been specifically defined as the system response which follows the initial response and precedes the steady-state response. In other words, it is the system response as it adjusts from its initial response to its steady state.

In referring to this adjustment, particularly with repetitive waveforms, it is convenient to use the term *transient effect* to describe a related effect which coincides with the transient response and the term *transient* to describe the difference between the steady-state and transient responses. Using this terminology the transient response becomes the steady state when the transient effects have ceased, or the transient is theoretically zero, or practically imperceptible.

The transient response of a system can frequently be atypical and possibly stress the system more than the steady-state response. This effect is demonstrated in some of the examples given in this chapter. Transient analysis can be used to quantify the magnitude and duration of extreme conditions where they occur. The extent of the transient effects may also be important in systems which are required to be switched from one state to another as rapidly as possible. How quickly the system settles down to the new state may need to be quantified.

Steady-state response
For the models of many systems the response never theoretically reaches a steady state or can be regarded to only have done so after infinite time. However, by displaying the response waveforms the observer can see how quickly any change actually becomes imperceptible. Some system models, or others operating under specific conditions, have no transient response and the initial response is followed immediately by the steady state.

Other applications exist in which systems are continually switched between one state and another. Examples of the models of such systems are circuits responding to pulse waveform sources, where the steady-state response may have the appearance of a succession of transient responses, as the voltage switches between different levels.

The strength of the computer as an analytical tool for displaying response waveforms is used throughout this chapter to provide a graphical appreciation of circuit behaviour. This is demonstrated with simple combinations of resistance, inductance and capacitance to illustrate significant aspects of circuit responses.

The chapter concludes with an introduction to state variable analysis which can be extended into a comprehensive general technique for producing the time-domain responses of more complex circuits.

5.1.1   Types of voltage source

A number of different source waveforms are considered in this chapter. Figure 5.1 illustrates the symbols used to indicate the type of waveform produced by the voltage source being considered. Both voltage and current sources are considered in the circuit analyses but the common practical realizations are voltage generators.

The salient features of the source types in relation to the analysis in this chapter are indicated below.

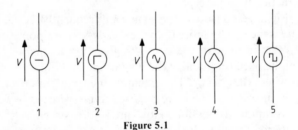

Figure 5.1

1. Steady value. The source maintains the steady voltage indicated between the two points in the circuit to which it is connected, from the instant it is applied.

2. Step voltage. The source voltage rises instantly to the value indicated at the time datum and then remains constant. This is equivalent to applying a steady voltage to the circuit at time zero.

3. Sine wave. The source voltage varies sinusoidally with time and has a specified repetition rate. In transient analysis the zero time angle for the sinusoidal source is important to indicate the initial value of the applied voltage.

4. Triangular waveforms. The source voltage rises and falls at a constant rate between specified voltage levels. The waveform has a specific repetition rate and in transient analysis the time datum is important.

5. Pulse waveforms. The source voltage rises and falls instantaneously between prescribed constant levels at specific times. Real pulse generators will have finite rise and fall times between the levels of constant voltage.

⟩ **5.2 Step input to inductance and capacitance**

A broad appraisal of how a circuit will respond to a sudden change can often be formed, without detailed analysis, by applying a fundamental understanding of component behaviour. Before considering more realistic combinations of elements it is therefore important to understand how the models of pure inductance and capacitance behave.

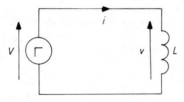

Figure 5.2

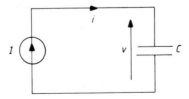

Figure 5.3

The circuit elements in figures 5.2 and 5.3 are supplied with step inputs of voltage and current, respectively. The analysis is performed in terms of unspecified values and then the graphs are demonstrated for unit step inputs of 1 V and 1 A.

The analysis for the two circuits is as follows.
For the circuit of figure 5.2,

$$V = L(\mathrm{d}i/\mathrm{d}t) \qquad \mathrm{d}i/\mathrm{d}t = V/L.$$

Since V/L is constant, $\mathrm{d}i/\mathrm{d}t$ is constant and the current/time graph is a straight line.

Initial values: if $i = 0$ at $t = 0_-$ then $i = 0$ at $t = 0_+$ because the current in an inductance cannot be changed instantaneously, since this would mean infinite $\mathrm{d}i/\mathrm{d}t$ and infinite voltage. Hence $i = (V/L)t$.

For the circuit of figure 5.3,

$$I = C(\mathrm{d}v/\mathrm{d}t) \qquad \mathrm{d}v/\mathrm{d}t = I/C.$$

Since I/C is constant, $\mathrm{d}v/\mathrm{d}t$ is constant and the voltage/time graph is a straight line.

Initial values: if $v = 0$ at $t = 0_-$ then $v = 0$ at $t = 0_+$ because the voltage drop across a capacitance cannot be changed instantaneously, since this would mean infinite $\mathrm{d}v/\mathrm{d}t$ and infinite current. Hence $v = (I/C)t$.

The responses are summarized in table 5.1.

Table 5.1 Summary of responses.

Circuit of figure 5.2	Circuit of figure 5.3
Initial response $t = 0_+$	
$i = 0$	$v = 0$
$v = V$	$i = I$
Transient response $t = 0_+ -$infinity	
$i = (V/L)t$	$v = (I/C)t$
$v = V$	$i = I$
Steady-state response $t = $ infinity	
$i = $ infinity	$v = $ infinity
$v = V$	$i = I$

Response display
The voltage and current graphs are as shown in figures 5.4 and 5.5 for unit step inputs of voltage and current, respectively. Program filename 'LC' is used to illustrate these circuits. The element values can be

changed to demonstrate their quantitative relationship to the rate of change of current or voltage for the unit step source.

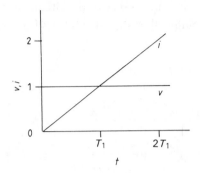

Figure 5.4 $v = 1$ V, $L = 1$ mH, $T_1 = 1$ ms.

As indicated above, the important property of the circuit elements when determining the initial response is that inductance current and capacitance voltage cannot be changed instantaneously. This means that the values of these parameters immediately before and immediately after any change in the system remain the same.

Unenergized inductance and capacitance thus behave instantaneously as open and short circuits, respectively. It follows therefore that an ideal current source cannot be applied to an inductance and an ideal voltage source cannot be applied to a capacitance. The models are incompatible, unless the elements already have the appropriate amounts of stored energy.

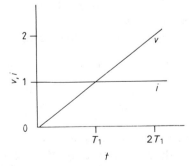

Figure 5.5 $v = 1$ V, $L = 1$ mH, $T_1 = 1$ ms.

⟩ 5.3   Step input to R–L and R–C circuits

The circuits in figures 5.6 and 5.7 are used to demonstrate the effects of adding resistance or conductance to the circuits in §5.2. For the analysis the circuits are supplied with step inputs of voltage V and current I respectively.

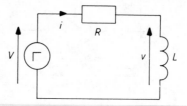

Figure 5.6

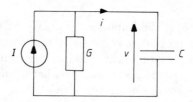

Figure 5.7

The analysis for the circuit of figure 5.6 is as follows:

$$V = Ri + v \qquad V = Ri + L(\mathrm{d}i/\mathrm{d}t) \qquad \mathrm{d}i/\mathrm{d}t = V/L - Ri/L.$$

Initial values: if $i = 0$ at $t = 0_-$ then $i = 0$ at $t = 0_+$.
Final values: $i = I_\mathrm{m}$ when $\mathrm{d}i/\mathrm{d}t = 0$, hence $I_\mathrm{m} = V/R$ and $\mathrm{d}i/\mathrm{d}t = (I_\mathrm{m} - i)/T$, where $T = L/R$.
For the circuit of figure 5.7 the following applies:

$$I = Gv + i \qquad I = Gv + C(\mathrm{d}v/\mathrm{d}t) \qquad \mathrm{d}v/\mathrm{d}t = I/C - Gv/C.$$

Initial values: if $v = 0$ at $t = 0_-$ then $v = 0$ at $t = 0_+$.
Final values: $v = V_\mathrm{m}$ when $\mathrm{d}v/\mathrm{d}t = 0$, hence $V_\mathrm{m} = I/G$ and $\mathrm{d}v/\mathrm{d}t = (V_\mathrm{m} - v)/T$ where $T = C/G$.
The significance of T is explained in §5.4.
For these circuits the rates of change of current and voltage are not

constant but proportional to the difference between the final and instantaneous values, $(I_m - i)$ and $(V_m - v)$, respectively. This is the property of the exponential curve and the differential equations can be solved by integration giving the following expressions:

$$\text{for figure 5.6} \quad i = I_m[1 - \exp(-t/T)]$$

$$\text{for figure 5.7} \quad v = V_m[1 - \exp(-t/T)].$$

The analysis required to produce the exponential functions is given in Appendix A5.3.

For figure 5.6, the inductance voltage is given by the equation $v = V - Ri = V[\exp(-t/T)]$. For figure 5.7, the capacitance current is given by the equation $i = I - Gv = I[\exp(-t/T)]$.

Alternatively the graphs of current and voltage against time can be obtained by linearized approximations, as shown in figures 5.10 and 5.11. The values of inductance current against time are given in table 5.4 to demonstrate the level of approximation and quantify the relationship to T.

Zero sources
After a time $t = 5T$ s, the voltage and current sources are reduced to zero and the circuits are then as shown in figures 5.8 and 5.9.

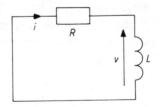

Figure 5.8

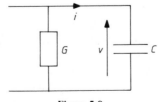

Figure 5.9

The following analysis applies to these circuits:

for figure 5.8: $0 = Ri + L(di/dt)$ $di/dt = -(R/L)i$

for figure 5.9: $0 = Gv + C(dv/dt)$ $dv/dt = -(G/C)v.$

For these circuits the current and voltage decrease at a rate proportional to the instantaneous values i and v. This is the property of the decreasing exponential curve and i and v decrease from their initial values at the instant the sources are reduced to zero to their final values of zero.

Values at $t = 5T$: for figure 5.8, at $t = 5T_-$,

$$i = I_m[1 - \exp(-5)]$$

hence $i = I_m$ approximately and $v = 0$ approximately.
For figure 5.9 at $t = 5T_-$,

$$v = V_m[1 - \exp(-5)]$$

hence $v = V_m$ approximately and $i = 0$ approximately.

Taking the approximate values to be the actual values and using a new time datum to simplify the expressions, i.e. $t(\text{new}) = t(\text{old}) - 5T$, then the following applies.

For figure 5.8, at $t = 0_+$, $i = I_m$ and $v = -RI_m = -V$, hence $i = I_m \exp(-t/T)$ and $v = -Ri = -V \exp(-t/T)$.

At the instant the voltage source is reduced to zero the inductance produces a voltage with the appropriate magnitude and polarity to maintain the current in the inductance and resistance.

For figure 5.9, at $t = 0_+$, $v = V_m$ and $i = -GV_m = -I$, hence $v = V_m \exp(-t/T)$ and $i = -Gv = -I \exp(-t/T)$.

At the instant the current source is reduced to zero the capacitance produces a current with the appropriate magnitude and direction to maintain the voltage across the capacitance and conductance.

The responses are summarized in tables 5.2 and 5.3.

Programs under filename 'TR' produce the graphs to illustrate the analyses in §§5.3 and 5.3.1 (see figures 5.10 and 5.11 for the $R-L$ and $R-C$ circuits respectively). The graphs are normalized for axes of maximum voltage and current and $10T$, for the time axis. The new time datum for the zero sources corresponds to $5T$. The element values are specified and quantify the scale factors of the axes for the graphs relating to specific circuits.

Table 5.2 Summary of responses.

Circuit of figure 5.6	Circuit of figure 5.7
Step input V	Step input I
Initial response $t = 0_+$	
$i = 0$	$v = 0$
$v = V$	$i = I$
Transient response $t = 0_+ -$infinity	
$i = I_m[1 - \exp(-t/T)]$	$v = V_m[1 - \exp(-t/T)]$
$v = V \exp(-t/T)$	$i = I \exp(-t/T)$
Steady-state response $t =$ infinity	
$i = I_m$	$v = V_m$
$v = 0$	$i = 0$

Table 5.3 Summary of responses.

Circuit of figure 5.8	Circuit of figure 5.9
Step reduction to zero voltage source at new time datum	Step reduction to zero current source at new time datum
Initial response $t = 0_+$	
$i = I_m$	$v = V_m$
$v = -V$	$i = -I$
Transient response $t = 0_+ -$infinity	
$i = I_m \exp(-t/T)$	$v = V_m \exp(-t/T)$
$v = -V \exp(-t/T)$	$i = -I \exp(-t/T)$
Steady-state response $t =$ infinity	
$i = 0$	$v = 0$
$v = 0$	$i = 0$
$T = L/R$	$T = C/G$
$I_m = V/R$	$V_m = I/G$

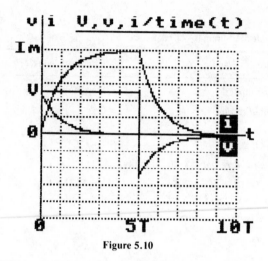

Figure 5.10

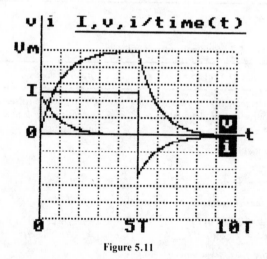

Figure 5.11

5.3.1   Dual circuits

The inclusion of resistance or conductance allows either a current or a voltage source to be applied to each circuit, as shown in figures 5.12 and 5.13.

These circuits demonstrate the duality of $R-L$ and $R-C$ circuits which generate the same relationships when voltage and current, R and G and L and C, are respectively interchanged.

The following analysis applies to these circuits.
For the circuit of figure 5.12,

$$I = Gv + i \qquad I = GL(di/dt) + i \qquad di/dt = (I - i)/T$$

where $T = LG$.

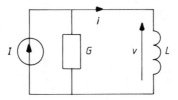

Figure 5.12

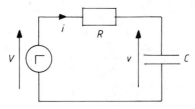

Figure 5.13

For the circuit of figure 5.13,

$$V = Ri + v \qquad V = RC(dv/dt) + v \qquad dv/dt = (V - v)/T$$

where $T = CR$. These equations are similar to those in §5.3, hence the following is obtained.

For figure 5.12, $i = I[1 - \exp(-t/T)]$ with the initial and final values of i being 0 and I, respectively.

The inductance voltage is given by the equation

$$v = (I - i)/G = I[\exp(-t/T)]/G.$$

For figure 5.13, $v = V[1 - \exp(-t/T)]$ with the initial and final values of v being 0 and V, respectively.

The capacitance current is given by the equation

$$i = (V - v)/R = V[\exp(-t/T)]/R.$$

The similarity of the expressions in §§5.3 and 5.3.1 is to be expected since, apart from the inductance and capacitance elements, the remainder of the circuit can be modelled as a Thevenin or a Norton equivalent generator. The voltage and current source circuits are equivalent with respect to the inductance and capacitance if $R = 1/G$ and $V = RI$.

Zero sources
After a time $t = 5T$ s, the current and voltage sources are reduced to zero, producing the circuits shown in figures 5.14 and 5.15. Note that a new time datum is used from the instant the sources are reduced to zero.

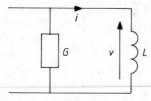

Figure 5.14

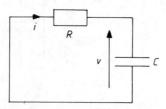

Figure 5.15

The following analysis applies to these circuits.
For figure 5.14,

$$0 = Gv = i \qquad 0 = GL(\mathrm{d}i/\mathrm{d}t) + i \qquad \mathrm{d}i/\mathrm{d}t = -i/T.$$

For figure 5.15,

$$0 = Ri + v \qquad 0 = RC(\mathrm{d}v/\mathrm{d}t) + v \qquad \mathrm{d}v/\mathrm{d}t = -v/T.$$

These equations are similar to those in §5.3, hence the following is obtained.
For figure 5.14,

$$i = I \exp(-t/T)$$

with the initial and final values of i being I and 0, respectively.
The inductance voltage is given by the equation

$$v = -i/G = -I[\exp(-t/T)]/G.$$

For figure 5.15,
$$v = V \exp(-t/T)$$

with the initial and final values of v being V and 0, respectively.
The capacitance current is given by the equation

$$i = -v/R = -V[\exp(-t/T)]/R.$$

The graphs produced by program filename 'TR' are shown in figures 5.16 and 5.17 for the *R–L* and *R–C* circuits respectively. The new time datum for the zero sources corresponds to $5T$.

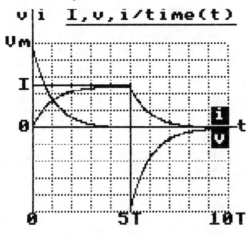

Figure 5.16

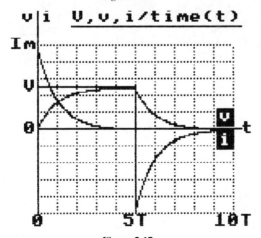

Figure 5.17

⟩ 5.4   Time constant and linearized approximation

5.4.1   Time constant

For the increasing or decreasing exponential curves the ratio

$$(\text{final value} - \text{present value})/\text{rate of change}$$

is a constant time referred to as the *time constant* of the system.

Table 5.4

Time (T)	Current increasing (I_m)	Current decreasing (I_m)
0.00	0.00	1.00
0.20	0.18	0.82
0.40	0.33	0.67
0.60	0.45	0.55
0.80	0.55	0.45
1.00	0.63	0.37
1.20	0.70	0.30
1.40	0.76 H	0.24 L
1.60	0.80	0.20
1.80	0.84 H	0.16 L
2.00	0.87 H	0.13 L
2.20	0.89	0.11
2.40	0.91	0.09
2.60	0.93	0.07
2.80	0.94	0.06
3.00	0.95	0.05
3.20	0.96	0.04
3.40	0.97	0.03
3.60	0.97	0.03
3.80	0.98	0.02
4.00	0.98	0.02
4.20	0.99	0.01
4.40	0.99	0.01
4.60	0.99	0.01
4.80	0.99	0.01
5.00	0.99	0.01

For the inductive circuit the time constant, T, is given by $T = (I_m - i)/(di/dt)$ or $T = -i/(di/dt)$, where $T = L/R$ or LG s.

For the capacitive circuit the time constant, T, is given by $T = (V_m - v)/(dv/dt)$ or $T = -v/(dv/dt)$, where $T = C/G$ or CR s.

After a time equal to one time constant the exponential curve has changed by approximately 63% and after a time of five time constants over 99% of the total variation between the initial and final values has occurred.

The values for the increasing and decreasing curves produced by program filename 'TR', are shown in table 5.4. The values are compared with the exponential function to indicate the accuracy of the algorithm used to produce the graphical responses. To the decimal places quoted, values marked with H are 0.01 high and those marked with L, 0.01 low.

5.4.2   Approximate linearized responses

The concepts of time constant and the principle of approximate linearized responses are closely related and are illustrated in figure 5.18.

 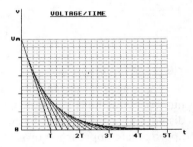

Figure 5.18

At values of current or voltage separated by a chosen time interval, the rate of change is calculated and the graph is then drawn as a straight line continuation with the calculated slope. The straight line produced in this process intersects the final value of the dependent parameter one time constant later.

This approximate method of obtaining graphical responses is based on the meaning of rate of change, namely

$$\text{rate of change} = \frac{\text{(new value} - \text{old value)}}{\text{time interval between values}}.$$

For graphs determined in this manner the linear segments are not extended to the final value but only to the point at which the slope is recalculated. The extension is included here to illustrate the principle of time constant.

Clearly the approximate solution approaches the actual response curve more closely as the time interval between successive slope calculations decreases. A compromise has therefore to be made between computing time and accuracy.

Program filename 'TC' produces the graphs in figures 5.18 and 5.19, showing increasing accuracy with reduction in time interval.

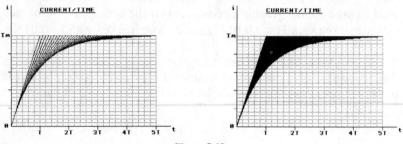

Figure 5.19

5.4.3   More accurate numerical techniques

It is not appropriate in this context to discuss in detail the various numerical methods for solving differential equations. For graphical displays aimed at producing an appreciation of circuit behaviour, the approximate approach shown in figures 5.18 and 5.19, with sufficiently small time intervals, may well be adequate.

In applications where greater accuracy is required more sophisticated methods might be necessary. The numerical method used must be properly chosen for its task and it is important to note that, whereas the local errors at each step may be small if they are all the same sign, the accumulated global errors may become unacceptably high.

The approach adopted here is to use the Euler predictor method for purely illustrative displays, with sufficiently small time intervals related to the circuit parameters. For more accurate displays and other problem-solving programs a predictor–corrector method with interval adjustment is used. This choice is appropriate since, in the case of the former, there is a close relationship between the numerical method and the circuit behaviour and in the case of the latter the process is similar to an error-correcting feedback system, commonly used in engineering prac-tice. The predictor–corrector algorithm used is described briefly in Appendix A5.4.3.

For more detailed information on numerical methods see the suggested texts in the complementary and further reading list.

〉 5.5 Transient and steady-state responses: repetitive waveforms

There are unlimited variations of circuits and repetitive waveforms. The examples given here are consequently restricted to illustrating specific circuit properties with commonly occurring waveforms.

The duality of $R-L$ and $R-C$ circuits has already been demonstrated and rather than duplicate the diagrams for all of the circuit conditions some are left as practice problems.

A major quality of the computer workstation as an analysis tool is its ability to produce function-driven graphics. This quality is exploited in the remainder of this chapter to demonstrate transient and steady-state behaviour.

Conventional analysis relies on functions to indicate circuit behaviour. By using computer graphics, circuit responses, whether as functions, or the result of numerical analyses, can be accurately displayed and the behaviour of the circuit observed directly. This visual interpretation of circuit behaviour allows for accompanying explanations to be relatively brief.

In some instances the waveforms are shown in the text with a grid background to facilitate quantitative interpretation. This is an option when running the programs. The instantaneous values of the response waveform are additionally displayed as the graphs are plotted. The plotting procedure can be interrupted at any instant to allow the values to be noted.

For each circuit the initial, transient or steady-state response is selected by specifying the highest cycle number (HC) of the response waveform displayed in relation to the number of cycles (NC) displayed, i.e. if HC = NC the initial response is displayed and if HC ≫ NC the steady state. The relative magnitudes of HC and NC in the latter case will depend on how quickly the transient response changes.

The repetitive waveforms used in the following sections are sine, triangular and pulse. The significant parameters for each source, which affect the transient and steady-state responses, are discussed at the beginning of the appropriate section. The response waveforms in all cases are determined for circuits with zero stored energy at the instant the source is applied, i.e. zero initial conditions at time 0_.

⟩ 5.6   Transient and steady-state responses: sine waves

5.6.1   General parameters
The waveform displays for the examples given in this section and for other similar problems are produced by programs under filename 'SNW', for which the introductory frame is shown in figure 5.20.

SINE-WAVE RESPONSES

This program draws initial, transient and steady-state responses of, series R-L and R-C, circuits for sine-wave voltage sources.

The wave parameters shown below can be varied.

The sensitivity and accuracy of the display can be increased by increasing the existing values when prompted.

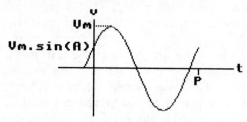

Figure 5.20

The frequency-domain analysis in Chapter 4 produces steady-state sine-wave responses in terms of impedance phasors, the magnitude of the phasor determining the amplitude of the current or voltage response and the angle of the phasor the phase difference between the appropriate voltage drop and current. The time-domain analysis in this section for producing both transient and steady-state responses models the circuit in terms of differential equations.

The value of the sinusoidal source at zero time does not affect steady-state phasor analysis, which is concerned with relative phase only.

In transient analysis the initial value of the source has a significant effect on the transient response and must be specified. For the sinusoidal voltage shown in figure 5.20 and given by $v = V_m \sin(\omega t + A)$, the value at zero time is $V_m \sin A$ and both V_m and A are input parameters for the following waveform displays, together with the period, $P(= 2\pi/\omega)$. The value of the phase angle, A, is quoted in degrees for each example.

One other relationship which contributes to the general framework in which the response waveforms are regarded is the relative magnitude of the circuit time constant (T) and the sine-wave period. The two aspects of this relationship which are considered are detailed below.

(1) The magnitude of the circuit time constant in relation to the sine-wave period determines for how many cycles the transient is significant, i.e. if $T \ll P$ the transient occurs within one cycle and over several cycles if $T \gg P$.

(2) The magnitude of the time constant in relation to the sine-wave period and the resulting steady-state phase difference between the source voltage and current can be quantified as follows.

The phase difference between voltage drop and current for the $R-L$ series circuit is $\tan^{-1}(\omega L/R)$ and $\tan^{-1}[1/(\omega CR)]$ for the series $R-C$ circuit. Expressed in terms of T and P, the phase differences are therefore $\tan^{-1}(2\pi T/P)$ and $\tan^{-1}[P/(2\pi T)]$, respectively. Hence for the $R-L$ circuit the phase difference is small if $T \ll P$ and tends to $90°$ if $T \gg P$. For the $R-C$ circuit the reverse is true and the phase difference is small if $T \gg P$ and tends to $90°$ if $T \ll P$.

The following waveforms demonstrate some of the important aspects of transient and steady-state sine-wave responses within the framework described above.

5.6.2 Sine-wave input to R–L and R–C circuits

(i) R–L circuit, P = 6.28T, Phase difference = 45°
The inductance current is zero at time 0_- and 0_+ producing a transient effect which is not visible after one cycle. The steady-state current wave lags the input voltage by $45°$ and the inductance voltage drop by $90°$. (See figure 5.21.)

(ii) R–C circuit, P = 6.28T, phase difference = 45°
The capacitance voltage drop is zero at time 0_- and 0_+ producing a

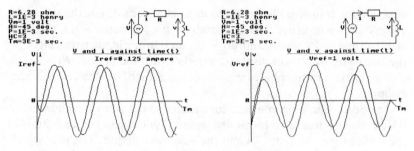

Figure 5.21

transient which is not visible after one cycle. The steady-state current leads the input voltage by 45° and the capacitance voltage drop by 90°. (See figure 5.22.)

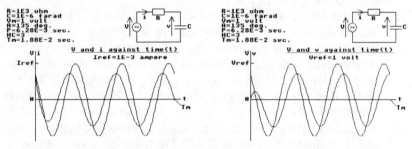

Figure 5.22

Zero and maximum transient conditions
Figure 5.23 illustrates the significance of the zero-time phase angle in relation to transient effects. The first three cycles are shown for voltage source zero-time angles of 89° and − 1°, respectively. For the 89° angle there are no transient effects and for the − 1° condition the transient and steady-state responses are markedly different.

In general the initial and steady-state responses are identical, producing no transient effects when the zero-time phase angle is related to the steady-state phase difference. This relationship must be such that the phase difference and circuit current and voltage at zero time are compatible with the steady-state phase difference and the circuit elements. This means that for the inductive circuit the voltage must be applied with the appropriate phase angle at an instant corresponding to a steady-state current zero since the inductance current cannot be changed instantaneously.

Hence the zero-time phase angle A for the zero transient condition is given by

$$A = \tan^{-1}(\omega L/R) = \tan^{-1}(2\pi T/P)$$

or

$$A = -[180 - \tan^{-1}(\omega L/R)] = -[180 - \tan^{-1}(2\pi T/P)]^{\circ}.$$

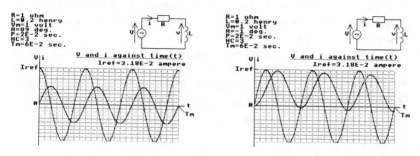

Figure 5.23

The initial and steady-state conditions are least compatible for the inductive circuit when the voltage is applied with the appropriate phase angle at an instant corresponding to a steady-state current maximum or minimum.

For the example shown in figure 5.23 where $T \gg P$, i.e. the transient is apparent for several cycles and the steady-state phase difference is approximately 90°, the maximum transient response current is approximately twice the steady-state current amplitude.

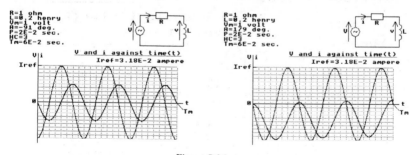

Figure 5.24

Figure 5.24 shows the alternative conditions for zero and maximum transients when the source voltages are phase shifted by 180° relative to those in figure 5.23.

Note that this example illustrates the difference between the zero and maximum transient conditions for a specific circuit. For other circuits in which T has a different relationship to P, the ratio

$$\frac{\text{maximum transient current}}{\text{steady-state current amplitude}}$$

may not be as great as in this example and the conditions required to produce the maximum value of the ratio may need more detailed investigation.

To produce the zero transient condition in the capacitive circuit the appropriate phase angle at the instant the voltage is applied must correspond with a steady-state capacitance voltage zero. The zero-time phase angle A for the zero transient condition is therefore given by

$$A = \tan^{-1}(\omega CR) = \tan^{-1}(2\pi T/P)$$

or

$$A = -[180 - \tan^{-1}(\omega CR)] = -[180 - \tan^{-1}(2\pi T/P)]^{\circ}.$$

Figure 5.25 illustrates one of the above zero transient conditions.

For both the inductive and capacitive circuits the relationships for the

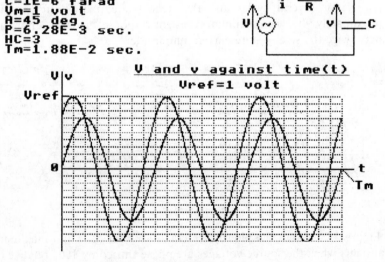

```
R=1E3 ohm
C=1E-6 farad
Vm=1 volt
A=45 deg.
P=6.28E-3 sec.
HC=3
Tm=1.88E-2 sec.
```

Figure 5.25

zero-time angles required to produce zero and maximum transient conditions will need to be modified if the inductance and capacitance have initial stored energy.

⟩ 5.7   Triangular-wave responses

5.7.1   General parameters

The waveform displays for the examples given in this section and for other similar problems are produced by programs under filename 'TRW', for which the introductory frame is shown in figure 5.26.

TRIANGULAR-WAVE RESPONSES

This program draws initial, transient and steady-state responses of, series R-L and R-C, circuits for triangular-wave voltage sources.

The wave parameters shown below can be varied.

The sensitivity and accuracy of the display can be increased by increasing the existing values when prompted.

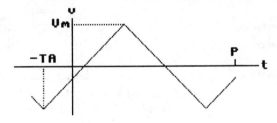

Figure 5.26

In a similar manner to the sine-wave responses discussed in the previous section, the value of the source waveform at zero time and the relative magnitudes of the waveform period (P) and the circuit time constant (T), will each affect the transient responses of circuits to triangular-wave source voltages.

There is, however, no corresponding steady-state theory with which to compare the responses produced in this section by differential equation models, although such comparison is possible by using the techniques of waveform synthesis described in Chapter 6.

5.7.2 *Triangular wave input to R–L and R–C circuits*

Triangular-wave responses in series $R–L$ and $R–C$ circuits are demonstrated by examples with the three extreme conditions detailed below.

(1) If most of the triangular source voltage is dropped across the resistance, the $v–i$ relationship of this element dominates the response, producing a triangular current waveform. For this condition, $T \ll P$ for the inductive circuit and $T \gg P$ for the capacitive circuit. This response is shown in figure 5.27.

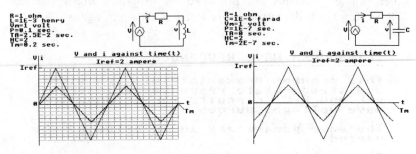

Figure 5.27

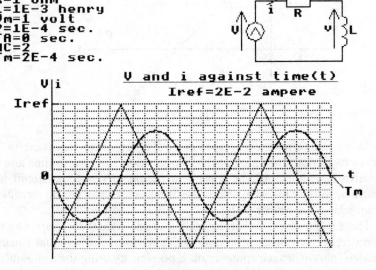

Figure 5.28

(2) The inductance $v-i$ relationship dominates the response for the $R-L$ circuit if most of the source voltage is dropped across it. A triangular voltage across an inductance is associated with a parabolic current response. This condition occurs when $T \gg P$ and is shown in figure 5.28.

(3) If the capacitance $v-i$ relationship dominates the response of the $R-C$ circuit, an approximate square wave of current is produced. For most of the source voltage to be dropped across the capacitance, $T \ll P$. This response is shown in figure 5.29.

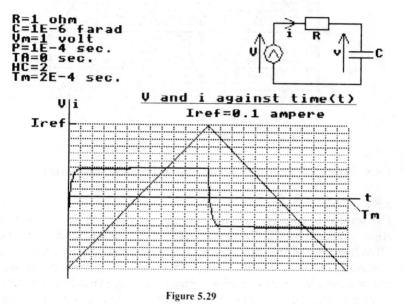

R=1 ohm
C=1E-6 farad
Vm=1 volt
P=1E-4 sec.
TA=0 sec.
HC=2
Tm=2E-4 sec.

V and i against time(t)
Iref=0.1 ampere

Figure 5.29

Minimum and maximum transient conditions
The responses shown in figures 5.28 and 5.30 demonstrate the effect of the value of the source voltage at zero time. These waveforms illustrate the transient effects for the conditions when zero time corresponds to a steady-state current zero and a current minimum respectively, i.e. most and least compatible with the zero-time requirements of the inductive circuit.

Note that the analyses in this section are descriptive only and based on relative magnitudes. Within the limits of display accuracy they are intended to give an appraisal of triangular-wave responses.

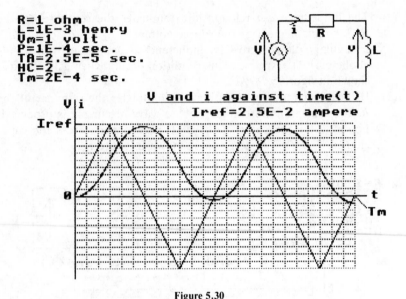

R=1 ohm
L=1E-3 henry
Vm=1 volt
P=1E-4 sec.
TA=2.5E-5 sec.
HC=2
Tm=2E-4 sec.

Figure 5.30

> **5.8   Transient and steady-state responses: pulse waveforms**

5.8.1   General parameters

The waveform displays for the examples given in this section and for other similar problems are produced by programs under filename 'PLW', for which the introductory frame is shown in figure 5.31.

Waveforms with rapid changes inherent in their shape produce responses similar to the transients which occur when circuits are switched on or off. The pulses considered here change from one constant voltage level to another in zero time. In mathematical terms they are discontinuous waveforms. In practice, there is always a finite rise or fall time but in many instances this is negligible when compared with the pulse period and can be neglected.

Figure 5.31 shows details of one duty cycle of a repetitive pulse train with different magnitudes of upper and lower voltage level and unequal positive and negative pulse widths.

The following relationships apply: pulse frequency $= f$ Hz, pulse period, $P = 1/f$ s, upper voltage level $=$ UVL V, lower voltage level $=$ LVL V, duty factor, $D = PP/P$, i.e. the ratio, duration of upper voltage level/pulse period.

PULSE WAVEFORM RESPONSES

This program draws initial, transient
and steady-state responses of, series
R-L, R-C and R-L-C, circuits for pulse
waveform voltage sources.

The wave parameters shown below can be
varied.

The sensitivity and accuracy of the
display can be increased by increasing
the existing values when prompted.

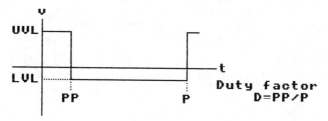

Figure 5.31

Square wave
The square wave is a specific pulse train for which

$$\text{LVL} = -\text{UVL and } D = 0.5.$$

5.8.2   Pulse input to R–L and R–C circuits

The difference between the various waveforms in this section derives
from the relationship between the circuit time constant and the pulse
parameters. The values of the circuit elements and pulse waveform
parameters have been chosen to illustrate a particular property and allow
easy confirmation of quantitative relationships, rather than to be typical
of actual circuits. The properties themselves are typical and scaling the
element values to an actual circuit will indicate its behaviour.

In general terms, it can be seen from the waveforms that, when the
time constant is much shorter than the pulse width, the transient decays
to zero within one pulse. Where the time constant is long compared with
the pulse width the transient persists for a number of pulses.

The initial, transient or steady-state response is selected, as described
generally for repetitive waveforms in §5.5, by the choice of the number
of cycles and the highest cycle number displayed. A more detailed
description of how the programs can be used is given in §5.10.

(i)   R–L circuit, P = 10T
The waveforms in figure 5.32 show that the current and voltage have almost reached their final values when the pulse level drops to zero. The current cannot fall instantaneously and the inductance voltage reverses to maintain it.

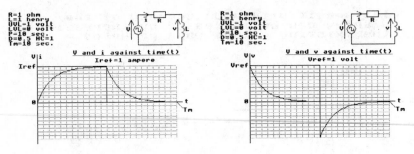

Figure 5.32

(ii)   R–C circuit, P = 10T
The capacitance voltage cannot fall instantaneously as the pulse level drops to zero and the current reverses as the capacitance discharges exponentially (see figure 5.33).

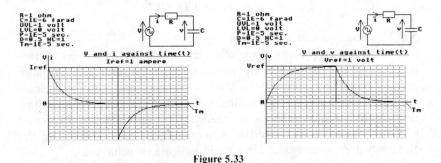

Figure 5.33

(iii)   R–L circuit, P = T
Figure 5.34 shows the average current waveform rising, over approximately five pulse periods, to a steady value of $(V_1/2)/R$, the average voltage divided by the resistance. The inductance voltage falls to an average value of zero, since any constant component of current produces no voltage drop across it.

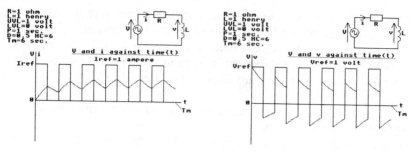

Figure 5.34

(iv)   Pulse–width modulation (PWM)

The waveforms in figure 5.35 illustrate the principle of using PWM to vary the average or DC level of a source and the current waveforms associated with the technique. The average level is proportional to the duty factor.

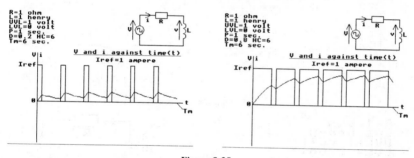

Figure 5.35

(v)   Steady-state current waveforms: square-wave input R–L and R–C circuits, P = 2T

Figure 5.36 shows the steady-state current waveforms for a square-wave

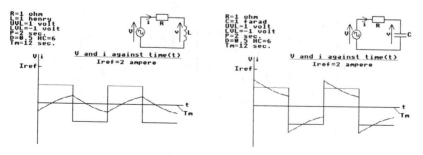

Figure 5.36

input to $R-L$ and $R-C$ series circuits. If two circuits with these specific, or similar sets of parameters, are connected in parallel, the source current for the resulting circuit is a square wave. This can be demonstrated by adding the current waveforms for the parallel branches. Circuits with this property are discussed in more detail in §5.9.2.

(vi)   Steady-state current and voltage waveforms: square-wave input R–L circuit, P = 10T

Figure 5.37 shows that the inductance voltage doubles at the instant the square wave changes polarity. In order to maintain the current, which cannot change instantaneously, the inductance voltage is required to

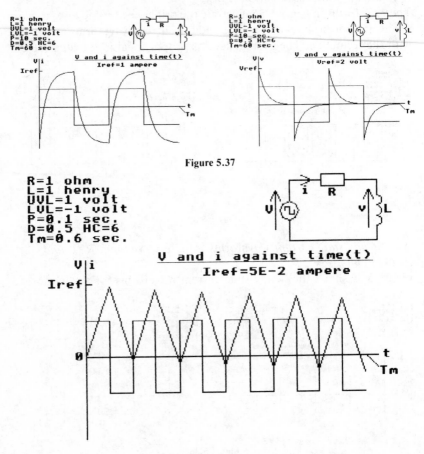

Figure 5.37

Figure 5.38

offset the negative square-wave voltage and produce an instantaneous supply equal to the positive voltage.

(vii)   R–L circuit, P = T/10
For this condition most of the source voltage is dropped across the inductance and thus the v–i relationship for this element dominates the response. The current waveform is therefore almost triangular (figure 5.38) and similar to the condition shown in figure 1.20. Considered in terms of the time constant, the pulse period is relatively short, hence only the initial part of the exponential curve occurs before the voltage polarity is reversed. The transient response changes relatively slowly over several cycles towards the steady-state response (see practice problem 6).

⟩ 5.9   Further aspects of time constant

The preceding sections in this chapter have demonstrated that the circuit time constant is an important parameter in relation to time-domain analysis. This is particularly evident with the relative magnitudes of the circuit time constant (T) and the waveform period (P) determining how rapidly any transient effect changes in relation to the repetitive waveform. Some further aspects of time constant are considered below.

5.9.1   Approximate differentiation and integration
The voltage–current relationships for inductance and capacitance given in §1.4 indicate that they are differential and integral for the pure elements. The response waveforms given in the preceding sections illustrate that, under certain conditions, the series combinations of resistance and inductance or capacitance behave as approximate integrating or differentiating circuits when the current is considered in relation to the source voltage. This means that for sinusoidal waveforms the phase difference must be approximately $90°$ and for triangular and square waveforms the current must be square, triangular, parabolic or impulsive as appropriate.

Some of the many possible examples of these effects are given below, with the current related to the source voltage.

(i)   In figures 5.23, 5.28 and 5.38 the current is proportional to the approximate integral of the voltage.

(ii)   In figure 5.29 the current is proportional to the approximate differential of the voltage.

For the above examples $T \gg P$ for the $R-L$ circuit for approximate integration and $T \ll P$ for the $R-C$ circuit for approximate differentiation. Alternative conditions to these exist if the inductance and capacitance voltage drop, instead of the current, are considered in relation to the source voltage (see practice problem 6).

This section has only covered differentiation and integration in approximate terms. Where accurate differentiation and integration are required more sophisticated circuits with active devices may become necessary.

5.9.2   Equal time constant circuit combinations
In this section the effects of combining two equal resistance, equal time constant circuits are considered. The circuits are connected first in parallel and then in series.

Equal time constant, equal resistance $R-L$ and $R-C$ circuits in parallel
If two equal time constant $R-L$ and $R-C$ circuits with equal resistances are connected in parallel as shown in figure 5.39 the overall circuit behaves as a pure resistance. This circuit has previously been analysed by frequency-domain methods in §4.9.3 where it was shown to behave as a pure resistance at all frequencies.

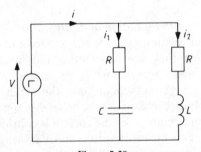

Figure 5.39

Considered as a time-domain problem with a step input to the circuit of V V, the capacitive branch current i_1 falls exponentially to zero from V/R, while the inductive branch carries a current i_2 which rises exponentially to V/R, both with the same time constant. The resulting source current i is therefore constant at V/R.

Using the expressions from §5.3

$$i = i_1 + i_2 = I_m \exp(-t/T) + I_m[1 - \exp(-t/T)] = I_m.$$

Alternatively, adding the two sets of figures obtained by computer analysis and given in table 5.4 demonstrates the same point.

The relationship between the circuit parameters for this condition is given by

$$L/R = CR$$
$$R = (L/C)^{1/2}.$$

Steady-state current waveforms for a square-wave input to equal time constant circuits are shown in figure 5.36.

This series–parallel counterpart of this circuit with a step current input is given as practice problem 8.

Equal time constant, equal resistance R–L and R–C circuits in series

For the series combination of a series $R–L$ and a series $R–C$ circuit with equal time constants and equal resistances, the analysis is more complex.

The series circuit produced is shown in figure 5.40 and the resistances

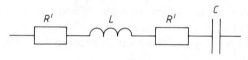

Figure 5.40

are shown combined in the circuit supplied with a step voltage source, shown in figure 5.41.

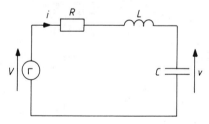

Figure 5.41

The relationships between the element values are

$$L/R' = CR' \qquad R' = (L/C)^{1/2}$$
$$R = 2R' \qquad R = 2(L/C)^{1/2}.$$

When the step source voltage is applied to the circuit, the inductance prevents an immediate rise in current and the capacitance produces a final current of zero when its voltage drop is equal to the source voltage. Thus the current rises from zero and then decreases back to zero. The analysis to produce the response between these limits is as follows:

$$V = Ri + L(di/dt) + v$$

$$di/dt = -(R/L)i - (1/L)v + V/L$$

$$i = C(dv/dt)$$

$$dv/dt = (1/C)i.$$

Initial values: if, at $t = 0_-$, $i = 0$ and $v = 0$ then at $t = 0_+$, $i = 0$, $v = 0$, $di/dt = V/L$ and $dv/dt = 0$.

Final values: $dv/dt = 0$ and $di/dt = 0$ when $v = V$.

The same numerical techniques can be used to solve these equations as those for the other circuits in this chapter. The difference in this instance is that two variables, i and v, are determined from their rates of change.

If a pulse waveform is applied to the circuit the analysis is similar to that for the step input. When the voltage level of the pulse waveform changes neither the circuit current nor the capacitance voltage can change instantaneously and these values become the initial conditions for the new voltage level.

The graphs of i and v against time for the series circuit are shown in figure 5.42.

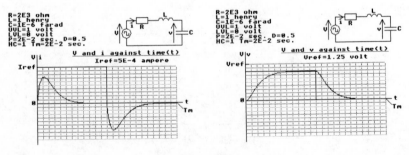

Figure 5.42

5.9.3 Unequal time constant circuits in series

If the resistance in the circuit in figure 5.41 is reduced from the equal time constant value then

$$R < 2(L/C)^{1/2}.$$

The response waveforms for this condition are shown in figure 5.43, where it can be seen that the current and capacitance voltage oscillate. For those parts of the response where the source voltage is positive and the current is negative the circuit is returning energy to the source.

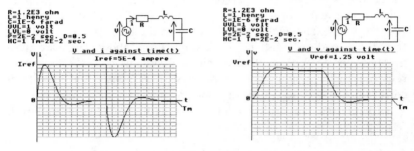

Figure 5.43

The third possibility for the series circuit is for the resistance to be increased from the equal time constant value, giving the expression

$$R > 2(L/C)^{1/2}.$$

The current and capacitance voltage waveforms for this condition are shown in figure 5.44.

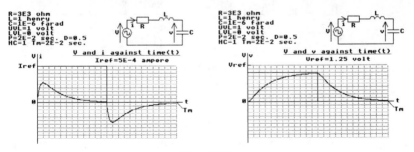

Figure 5.44

5.9.4   General R–L–C series circuits

Although the analysis of the series circuit in figure 5.41 has been approached from the combination of equal time constant circuits, the resistance clearly need not be associated equally with the inductance and the capacitance. The three types of response waveforms therefore apply to any series $R-L-C$ circuit where (i) $R = 2(L/C)^{1/2}$; (ii) $R < 2(L/C)^{1/2}$; (iii) $R > 2(L/C)^{1/2}$.

In terms of the Q factor for the series resonant circuit these conditions correspond to $Q = \frac{1}{2}$, $Q > \frac{1}{2}$ and $Q < \frac{1}{2}$, respectively.

It can be seen from figures 5.42, 5.43 and 5.44 that the response waveforms, which have been graphed to the same scale for comparison, reach their steady-state values in the shortest time for the condition when $R = 2(L/C)^{1/2}$. The analytical solutions for these circuits are produced from the second-order differential equations for i and v. These are given in Appendix A5.9. A number of physical systems have these types of response and analogous conditions. System responses and conditions corresponding to the waveforms shown in figures 5.42–5.44 are frequently referred to as critically damped, underdamped and overdamped, respectively.

5.9.5 Square-wave input to $R-L-C$ circuits
Figure 5.45 shows typical square-wave responses for the series $R-L-C$ circuit where $R < 2(L/C)^{1/2}$.

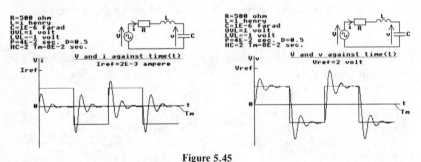

Figure 5.45

⟩ 5.10   Use of programs

The programs associated with this chapter fall into three broad groups designed for (i) simple step input response displays, (ii) the analysis and display of repetitive waveform responses and (iii) the more accurate step input analysis for circuits with initial conditions. The programs are listed in detail in Appendix B.

5.10.1   Repetitive waveform analysis and display
In the preceding sections, a number of examples have been given of the

use of the programs associated with this chapter for analysing and displaying the responses associated with repetitive waveforms. The program features previously described in the appropriate sections are summarized below.

Input parameters
The following parameters which affect the circuit response and hence the waveform display can be varied:

(*a*)   the circuit element values;
(*b*)   the amplitude and period of the input waveform;
(*c*)   the input waveform type: sine, triangular or pulse;
(*d*)   the zero time angle for sine-wave sources;
(*e*)   the time advance for triangular waveforms;
(*f*)   the duty factor for a pulse waveform;
(*g*)   the number of cycles displayed (up to six); and
(*h*)   the highest cycle number displayed.

Transient or steady state
The display window can be chosen to give either the transient or steady-state responses. If the transient response is required the first and successive cycles are displayed until the transient is imperceptible. The steady-state response is displayed by selecting a display window for a sufficiently high cycle number, by which time the transient has decreased to a negligible level. The program then computes the initial response but does not display those cycle numbers below the selected value.

Magnitude of voltage or current at any instant
The programs have an optional grid background for the display. This provides the facility for estimating voltage or current values with display level accuracy similar to an oscilloscope. The programs are also provided with a stop–start mechanism which enables the user to interrupt the display and read out the values at that instant.

Accuracy
When quoting response values the intended use and hence accuracy of the program must be considered. If the program is allowed to run for many computations, global errors may have accumulated to a significant magnitude.

A variable accuracy option is included in the programs. The default value, particularly for repetitive waveform displays, is designed to enable the user to obtain an estimate of the waveform, before setting the plotting scale and accuracy for a more precise display.

Note that the default accuracy is not intended to produce an adequate display.

Interpreting the accuracy of measured data is a general skill required of engineers; it must be similarly applied to computer methods. Accuracy is considered further in §6.6.

5.10.2   Circuit analysis example

Step input responses: system parameters
Analyses for the example in this section and for other similar problems are performed by programs under filename 'STI', for which the introductory frame is shown in figure 5.46. This example is based on the circuit shown in figure 5.47. The switch, S, in the circuit is closed for a specified time and then opened. The current response is required for two periods of time: after the switch is closed but before it is subsequently opened and after it is opened.

The solution can be divided into two parts:

(i)   the determination of the Thevenin equivalent for the circuit with respect to the inductance; and

```
STEP INPUT RESPONSES

This program draws initial, transient
and steady-state responses of, series
R-L,  R-C and  R-L-C, circuits for step
voltage sources.

The system parameters shown below can be
varied.

The sensitivity and accuracy of the
display can be increased by increasing
the existing values when prompted.


SYSTEM PARAMETERS

Element values R(ohm),L(henry),C(farad)
Step voltage V(volt)
Initial inductance current ILO(ampere)
Initial capacitance voltage VCO(volt)
Time-scale(second)
```

Figure 5.46

(ii) using program filename 'STI' with the modified circuit to determine the required values of current and voltage.

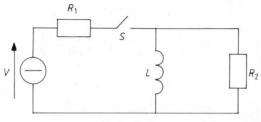

Figure 5.47

Thevenin equivalent

The Thevenin equivalent circuit with the inductance added is shown in figure 5.48, where

$$V_1 = [R_2/(R_1 + R_2)] V$$

$$R_3 = R_1 R_2/(R_1 + R_2).$$

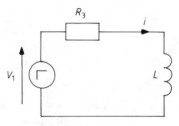

Figure 5.48

Step voltage analysis

When the switch is closed the inductance current rises exponentially with a time constant L/R_3. The initial value of the inductance current, I_0, immediately after the switch is opened, is unchanged from the value just before it is opened. To maintain this current through the inductance with the source disconnected, the current must flow through resistance R_2. The inductance voltage must therefore rise momentarily to $R_2 I_0$ before decreasing exponentially to zero with a time constant L/R_2. The inductance current decreases exponentially to zero from I_0 with a time constant L/R_2.

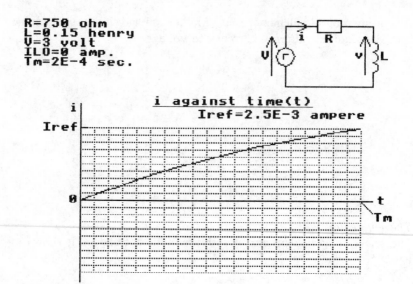

```
R=750 ohm
L=0.15 henry
V=3 volt
ILO=0 amp.
Tm=2E-4 sec.
```

Figure 5.49

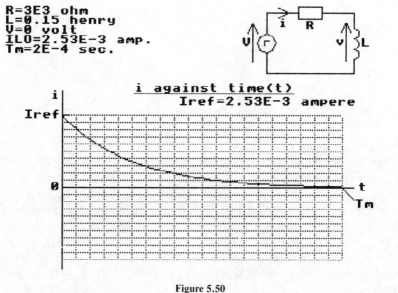

```
R=3E3 ohm
L=0.15 henry
V=0 volt
ILO=2.53E-3 amp.
Tm=2E-4 sec.
```

Figure 5.50

The values for the specific problem are as follows: $V = 4$ V, $R_1 = 10^3 \, \Omega$, $R_2 = 3 \times 10^3 \, \Omega$, $L = 0.15$ H, the switch is closed for 0.2×10^{-3} s and then opened.

The results of the computation are shown in figures 5.49 and 5.50. The timescale of the two graphs is the same, emphasizing the current decreasing with a shorter time constant. The inductance current at the instant the switch is opened is 2.53×10^{-3} A. This is the maximum value of current for the graph in figure 5.50. Note that for the graph in figure 5.50, $t = 0$ is a new time datum corresponding to the instant the switch is opened.

⟩ **5.11   State variable analysis**

The simple circuit structures which have been analysed in this chapter have been used to illustrate the types of transient responses associated with $R-L-C$ circuits. For the analysis of more complicated circuits general analytical tools are required. These are typically Laplace transform or complex variable techniques for functional analysis and numerical methods based on state variables.

The analysis of the series $R-L-C$ circuit in §5.9.2 produced two first-order differential equations for the inductance current, i, and the capacitance voltage, v. The variables i and v are referred to as the state variables of the circuit. State variable analysis consists of producing a set of simultaneous first-order differential equations with one equation for each inductance current or capacitance voltage. These equations can then be solved by one of a number of numerical techniques such as those described in previous sections. The state equations for circuits with a small number of state variables can be determined relatively easily by inspection but, as the number increases, more systematic methods become necessary.

The techniques are introduced and demonstrated here for circuits with one and then two state variables.

Program format
The general format of the program for state variable analysis, filename 'SV', is illustrated by the introductory frame shown in figure 5.51.

5.11.1   Circuits with one state variable
The state variable for the circuit shown in figure 5.52 is the inductance current i.

STATE VARIABLE ANALYSIS

This program computes the state variable time domain responses for circuits with step inputs and ONE or TWO state variables.

The state equations must be expressed in the general form,

ds1/dt = A.s1 + B.s2 + E

ds2/dt = C.s1 + D.s2 + F

where, s1 and s2, are the state variables and A, B, C, D, E and F, are constants.

Figure 5.51

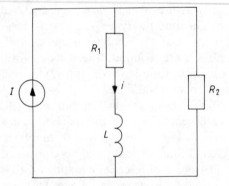

Figure 5.52

The state equation can be determined by applying Kirchhoff's current and voltage laws to the circuit as follows:

$$R_1 i + L \, di/dt = R_2(I - i)$$

hence

$$di/dt = - [(R_1 + R_2)/L]i + R_2 I/L.$$

For the circuit shown in figure 5.53 the state variable is the capacitance voltage and the state equation can be determined as follows:

$$V = R_1[C \, dv/dt + (v/R_2)] + v$$

hence

$$dv/dt = - [1/(CR_1) + 1/(CR_2)]v + V/(CR_1).$$

From these examples it can be seen that the state equation for circuits with one state variable has the general form

$$\mathrm{d}s/\mathrm{d}t = As + B \qquad (5.1)$$

where s is the state variable and A and B are constants.

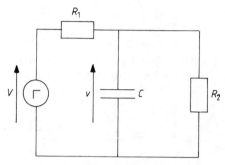

Figure 5.53

5.11.2   Single-state-variable example

The program under filename 'SV' for solving the general state equation (5.1) is demonstrated by using specific values for the circuit in figure 5.53 as follows: $R_1 = 1000 \ \Omega$, $R_2 = 1000 \ \Omega$ and $C = 10^{-6}$ F.

A step voltage of 1 V is applied to the circuit at time $t = 0$ s, when the voltage across the capacitance is $v = -0.5$ V. The subsequent time-domain response for the capacitance voltage is required.

The program requires the data in the form of the two constants A and

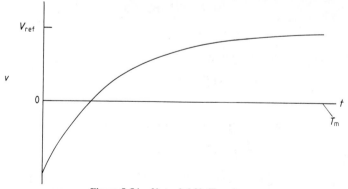

Figure 5.54 $V_{ref} = 0.5$ V, $T_m = 2$ ms.

B and the initial value of the voltage or current. Substituting the circuit element values in the equations given in §5.11.1 produces

$$A = -2 \times 10^3 \qquad B = 10^3.$$

The initial capacitance voltage is -0.5 V. The result of the computation is shown in figure 5.54.

5.11.3   Circuits with two state variables

Figure 5.55 shows a circuit with two state variables, the inductance current i and the capacitance voltage v, respectively. The two state equations can be determined by applying KCL and KVL to the circuit as shown below.

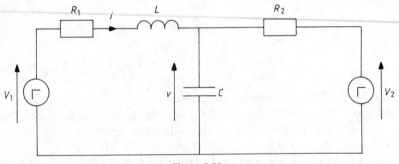

Figure 5.55

Applying KVL to the two meshes and KCL at the capacitance terminals produces the two equations

$$V_1 = R_1 i + L \, \mathrm{d}i/\mathrm{d}t + v$$

$$V_2 = R_2(C \, \mathrm{d}v/\mathrm{d}t - i) + v.$$

These equations can be rearranged to give

$$\mathrm{d}i/\mathrm{d}t = -(R_1/L)i - (1/L)v + V_1/L$$

and

$$\mathrm{d}v/\mathrm{d}t = (1/C)i - 1/(CR_2)v + V_2/(CR_2)$$

which can be expressed in the matrix form

$$\frac{\mathrm{d}}{\mathrm{d}t}\begin{bmatrix} i \\ v \end{bmatrix} = \begin{bmatrix} -R_1/L & -1/L \\ 1/C & -1/(CR_2) \end{bmatrix}\begin{bmatrix} i \\ v \end{bmatrix} + \begin{bmatrix} V_1/L \\ V_2/(CR_2) \end{bmatrix}.$$

The general matrix equation for circuits with two state variables is

$$\frac{d}{dt}\begin{bmatrix} s_1 \\ s_2 \end{bmatrix} = \begin{bmatrix} A & B \\ C & D \end{bmatrix}\begin{bmatrix} s_1 \\ s_2 \end{bmatrix} + \begin{bmatrix} E \\ F \end{bmatrix} \qquad (5.2)$$

where s_1 and s_2 are the two state variables and A, B, C, D, E and F are constants determined by the circuit parameters.

5.11.4 A circuit example with two state variables
The program under filename 'SV' for solving the general matrix state equation (5.2) is demonstrated on the circuit shown in figure 5.56.

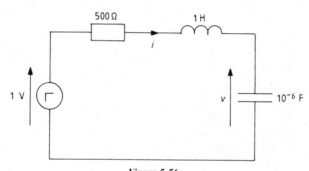

Figure 5.56

The series circuit shown in figure 5.56 has two state variables, the inductance current i and the capacitance voltage v.

Substituting the circuit parameters in the two state equations for the series $R-L-C$ circuit given in §5.9.2 produces

$$di/dt = -500i - v + 1$$

$$dv/dt = 10^6 i.$$

Hence the constants for the general expression (5.2) are $A = -500$, $B = -1$, $C = 10^6$, $D = 0$, $E = 1$ and $F = 0$.

The step voltage is applied at time $t = 0$ when the inductance current and the capacitance voltage are assumed to be zero. The results of the computation are shown in figures 5.57 and 5.58, s_1 being the inductance current and s_2 the capacitance voltage.

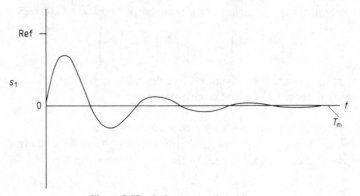

Figure 5.57   Ref = 1 mA, $T_m = 20$ ms.

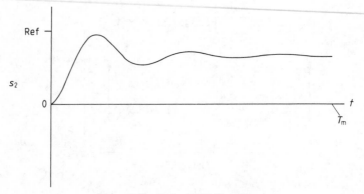

Figure 5.58   Ref = 1.5 V, $T_m = 20$ ms.

5.11.5   More general state variable analysis

Time-varying sources
The state variable analysis techniques in the previous sections are illustrated for circuits with sources which produce step functions of current or voltage. These sources change instantaneously at time $t = 0$ and then remain constant at the specified value. More generally, the sources may vary with time in which case the parameters E and F in equation (5.2) become in turn functions of time. The more general matrix form of the state equations is therefore

$$\frac{d}{dt}\begin{bmatrix} s_1 \\ s_2 \end{bmatrix} = \begin{bmatrix} A & B \\ C & D \end{bmatrix}\begin{bmatrix} s_1 \\ s_2 \end{bmatrix} + \begin{bmatrix} E(t) \\ F(t) \end{bmatrix} . \tag{5.3}$$

Output equations
Where the output response required is not a state variable, additional equations are required which express the output in terms of the state variables. For the example given in §5.11.4, if the required output response was the voltage drop across the resistance, v_R, then the output equation would be

$$v_R = 500i.$$

Neither of the above developments changes the technique in principle but both require extensions to any more general problem-solving programs. For a more advanced general treatment of state variable analysis consult the suggested texts given in the complementary and further reading list.

⟩ **5.12   Practice problems**

1   Compare the accuracy of programs under filenames 'PLW' and 'STI', for the circuits in §5.3 with the exponential expressions.

2   For the inductance circuit shown in figure 5.59 the switch is closed for 0.02 s and then opened;

   (i)   determine the current just before the switch is opened;
   (ii) determine the voltage across the 400 Ω resistance just after the switch is opened,
   (iii) display the inductance current and voltage waveforms from 0–0.04 s.

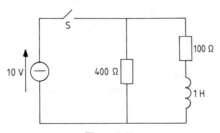

Figure 5.59

3   For the circuit shown in figure 5.60, $R = 10^5 \,\Omega$, $C = 10^{-6}\,\mathrm{F}$ and $V = 10\,\mathrm{V}$. The switch is closed at time $t = 0$; display the graph of capacitance voltage against time for $t = 0$–0.25 s.

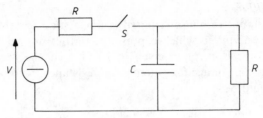

Figure 5.60

4   For the circuits shown in figure 5.61, the sinusoidal voltage source is applied at time $t = 0$ and the instantaneous value is given by $V = \sin(62830t + A)$ V. $R = 1000\ \Omega$, $L = 1$ H   and $C = 1\ \mu$F. Determine for each circuit:

(i)   the time constant (T) and the ratio T/periodic time (P);
(ii) the value of A in degrees for the minimum and maximum transient conditions.

Display the first two cycles of the i and v waveforms for the minimum and maximum transient conditions and determine for each case the ratio maximum transient value/steady-state amplitude.

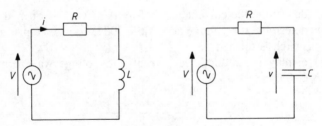

Figure 5.61

5   Display the current and voltage response waveforms produced when a triangular wave voltage source with zero time delay is applied to a series $R-L$ circuit. Consider the three conditions when the ratio time constant (T)/period (P) is given by (i) $T/P = 0.1$; (ii) $T/P = 1$; (iii) $T/P = 10$. Assume the maximum voltage is 1 V, the frequency is 1000 Hz and the resistance is 1 Ω.

Repeat the example for a series $R-C$ circuit, operating under the same conditions. Identify any approximate differential or integral relationships between the source and parameter waveforms.

6 Extend the material given in §§5.8 and 5.9 with the following practice problems.

(i) Display the current waveforms shown in figure 5.36 with a grid background. Sum the currents to determine the source current for a circuit consisting of the $R-L$ and $R-C$ branches, connected in parallel.

(ii) Using a resistance of 1000 Ω and a capacitance of 1 μF, display the equivalent waveforms for the $R-C$ circuit for the condition where $P = 10T$, shown for the $R-L$ circuit in figure 5.37.

(iii) For the circuit conditions shown in figure 5.38 estimate the lowest cycle number which approximates to the steady-state response. Display six cycles with a grid background to confirm the estimate.

(iv) Investigate the approximate differentiating and integrating properties of the series, $R-L$ and $R-C$ circuits if the voltage drop across the inductance or the capacitance is considered in relation to the source voltage. Relate T to P for each condition considered.

(v) By taking increased resistance values for the circuit shown in figure 5.45, display the response waveforms for the conditions where $R = 2(L/C)^{1/2}$ and $R > 2(L/C)^{1/2}$.

7 The circuit shown in figure 5.62 represents a simulation of a small permanent magnet motor turning at a steady speed, the steady 8 V source representing the motor EMF and the circuit elements the armature parameters. The motor speed is controlled by a PWM supply switched between 0 and 24 V at 1 kHz. Produce the steady-state current waveform for a duty factor of 0.5 and estimate the average current. Show the effect of changing the switching frequency to 500 Hz and 2 kHz. $R = 1.5\ \Omega$, $L = 2.3 \times 10^{-3}$ H.

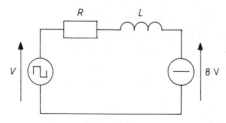

Figure 5.62

8   The circuit shown in figure 5.63 is supplied with a step current of
     I A. Discuss how the circuit behaves given that $LG = C/G$.

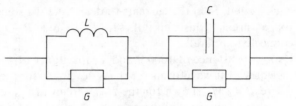

Figure 5.63

9   In the circuit shown in figure 5.64 the switch S is closed for
     5×10^{-3} s and then opened.

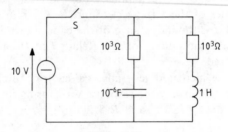

Figure 5.64

(i) Determine the following for the period during which the switch
     is closed:
     (i)   a graph of the inductance current against time,
     (ii) a graph of the capacitance voltage,
     (iii) a graph of the capacitance current,
     (iv) an expression for the source current.
(ii) Determine the following for the period after the switch is
     opened:
     (i)   a graph of the inductance current,
     (ii) a graph of the capacitance voltage,
     (iii) the maximum value of the inductance current,
     (iv) the time at which the maximum current occurs,
     (v)   the maximum value of inductance voltage indicating its
           polarity.

10   Apply program filename 'STI' to investigate the transient current

response in a series $R-L-C$ circuit with a step voltage source of 1 V. Use element values of 100–3000 Ω, 1 H and 1 μF. Where appropriate:

(i)   determine the frequency of oscillation of the current wave;
(ii) relate the resistance value to the rate at which the current oscillations decay;
(iii) compare the results of (a) and (b) with the resonant frequency and Q factor of the series $R-L-C$ circuit.

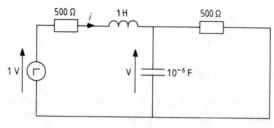

Figure 5.65

11   Display the transient response of the state variables i and v for the circuit shown in figure 5.65, assuming zero stored energy. State the final values of i and v.

〉 Chapter 6

〉 Waveform Synthesis and Analysis

〉 6.1   Overview

This chapter introduces subject matter related to general repetitive waveforms and draws together material from other parts of the book.

The process of waveform synthesis is demonstrated for a number of waveforms to convey the relationship between wave shape and harmonic content. This in turn illustrates the impact that frequency-sensitive elements and circuits have on voltage and current waveforms. Waveform analysis is introduced to demonstrate how the harmonic content of a repetitive wave can be determined. The processes of analysis and synthesis are then combined to show how the frequency-domain analysis given in Chapter 4, for pure sine waves, can be used to provide an alternative method for obtaining time-domain steady-state waveforms to that shown in Chapter 5.

The subject of non-linear analysis is introduced and the application of waveform analysis and synthesis techniques to the complex waveforms produced by non-linear circuit components is demonstrated.

The chapter ends with some general conclusions on the book as a whole.

〉 6.2   Waveform synthesis

In Chapter 3 the addition of voltage and current sine waves of the same frequency was considered. If, however, sine waves whose frequencies are integer multiples of a fundamental frequency are added, then repetitive

196

waveforms are produced which are non-sinusoidal. Such waveforms are described as complex waves.

The component frequencies of a complex wave are generally referred to as harmonics. The frequency of the complex waveform is equal to that of the first harmonic, referred to here and throughout this chapter as the fundamental. In the general case the complex wave may have an average or DC value and the harmonics different time data. Any DC component in the waveform simply moves the whole wave along the voltage or current axis and does not affect its shape.

The general series for a repetitive wave with a period of P s is given by

$$v = V_0 + V_1 \sin(\omega t + A_1) + V_2 \sin(2\omega t + A_2) + V_3 \sin(3\omega t + A_3) + \dots$$

where

$\omega = 2\pi/P$ rad s^{-1}, V_0 is the average value of the waveform, $V_1 \sin(\omega t + A_1)$ the fundamental, $V_2 \sin(2\omega t + A_2)$ the second harmonic and $V_3 \sin(3\omega t + A_3)$ the third harmonic, etc.

Note that V_1, V_2, V_3, etc are the *amplitudes* of the harmonics. The amplitude of a sine wave is its maximum value, as shown in figure 3.1. Where phasors are used to represent sine waves the magnitude of the phasor may be equal to the amplitude or the RMS value depending on the context.

Program filename 'SYNTH' is used to synthesize the waveforms shown in this chapter. The program can be used generally to explore the various aspects of waveform symmetry and to synthesize steady-state response waveforms in circuits with complex wave sources. The amplitude and phase of each component frequency of the complex wave is specified and then the effect of adding it to the existing waveform is shown. One cycle of the complex wave either side of the time datum is displayed from $-P$ to $+P$ s. This is done to illustrate any symmetry between the two cycles of the wave expressed as functions of time ($f(-t)$ and $f(t)$), either side of the time datum. The waveforms can be displayed against an optional grid background to facilitate some estimate of magnitudes.

6.2.1 Harmonic content

The harmonic content of a complex waveform can be quantified in terms of the harmonic and fundamental amplitudes.

Figure 6.1 shows the waveform which is produced when second and third harmonics are added to the fundamental, where the harmonic amplitudes are 50 and 30 %, respectively, of the fundamental amplitude.

The expression for the instantaneous voltage of this waveform is

$$v = 100 \sin \omega t + 50 \sin(2\omega t + \tfrac{1}{6}\pi) + 30 \sin (3\omega t + \tfrac{1}{3}\pi)$$

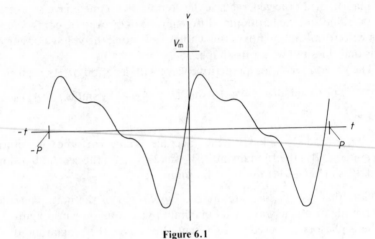

Figure 6.1

Time datum

Changing the time datum for the complex waveform changes the phase angles of the harmonics but not the amplitudes. Figure 6.2 shows the

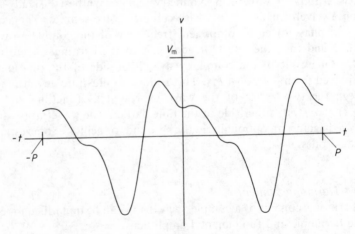

Figure 6.2

synthesis for a wave with the same harmonic content as that shown in figure 6.1 with the phase of the fundamental, second harmonic and third harmonic advanced by 90, 180 and 270°, respectively.

The equation for the time-shifted waveform is

$$v = 100 \sin(\omega t + \tfrac{1}{2}\pi) + 50 \sin(2\omega t - \tfrac{5}{6}\pi) + 30 \sin(3\omega t - \tfrac{1}{6}\pi).$$

The method of quantifying harmonic content as a percentage of the fundamental amplitude cannot be applied to waveforms in which the fundamental amplitude is zero. The waveform shown in figure 6.3 has the second and third harmonic components only from the waveform in figure 6.1 but its repetition frequency is still that of the zero-amplitude fundamental.

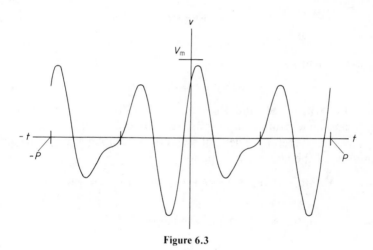

Figure 6.3

Line spectrum
Another method of representing the harmonic content of a complex waveform is by a line spectrum. This is a graph in which the harmonic amplitudes are represented by vertical lines scaled to the harmonic amplitude. The line spectra for the waveforms shown in figures 6.1 and 6.3 are shown in figure 6.4.

6.2.2 Sine and cosine series
It was shown in §4.2.4, in the context of phasor components, that a voltage represented by $V_1 \sin(\omega t + A_1)$ can be expressed as the sum of

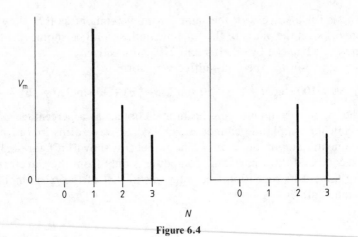

N

Figure 6.4

two component voltages,

$$V_{1s} \sin \omega t + V_{1c} \cos \omega t$$

where $V_{1s} = V_1 \cos A_1$ and $V_{1c} = V_1 \sin A_1$, V_{1s} and V_{1c} being the reference and quadrature components, respectively.

Using this property the general series for a complex waveform can be expressed by the alternative equation

$$v = V_0 + V_{1s} \sin \omega t + V_{2s} \sin 2\omega t + V_{3s} \sin 3\omega t + ... + V_{1c} \cos \omega t$$
$$+ V_{2c} \cos 2\omega t + V_{3c} \cos 3\omega t....$$

This is the Fourier series for the voltage v.

6.2.3   Waveform symmetry
Waveform symmetry can be used to give a broad appraisal of the harmonic content of a complex wave. This section illustrates some of the general aspects of waveform symmetry with examples of wave shapes with a limited number of harmonic components. The expression for the instantaneous voltage of each waveform is given as appropriate.

Odd functions
Figure 6.5 shows the effect of reducing the phase angle of the second and third harmonics of the waveform shown in figure 6.1 to zero. Such a series consisting only of sine waves produces a function $v(t)$ which

in mathematical terms is referred to as an odd function and can be expressed by the equation

$$v(t) = -v(-t)$$

i.e. the wave for negative values of time t is the inverted reflection, by the voltage or current axis through $t = 0$, of the wave for positive values of t.

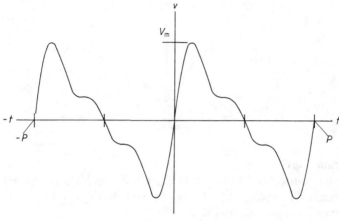

Figure 6.5

For an odd waveform the two half cycles are relatively inverted with a reversed time sequence, expressed for $t = 0 - \frac{1}{2}P$, by the equation

$$v(P - t) = -v(t).$$

Figure 6.5 shows an odd function for which the instantaneous voltage is given by

$$v = 100 \sin \omega t + 50 \sin 2\omega t + 33 \sin 3\omega t.$$

Sawtooth waveform
The waveform shown in figure 6.6 is an odd function produced by summing the fundamental and consecutive harmonics, the amplitude of which is inversely proportional to the harmonic frequency. The series for the odd function sawtooth waveform is given by

$$v = V \sin \omega t + \tfrac{1}{2} V \sin 2\omega t + \tfrac{1}{3} V \sin 3\omega t + \ldots (V/n) \sin n\omega t \ldots$$

and theoretically requires an infinite number of harmonics for its synthesis. The waveform shown in figure 6.6 contains the first twenty harmonics, with $V = 1$.

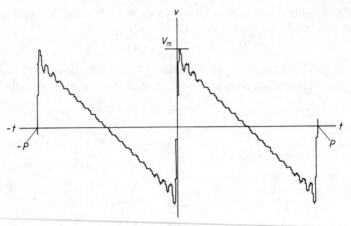

Figure 6.6

Even functions
Figure 6.7 illustrates the corresponding effect of adding cosine waves. A series consisting exclusively of cosine waves produces an even function which can be expressed by the equation

$$v(t) = v(-t)$$

i.e. the wave for negative values of time t is the reflection, by the voltage or current axis through $t = 0$, of the wave for positive values of t.

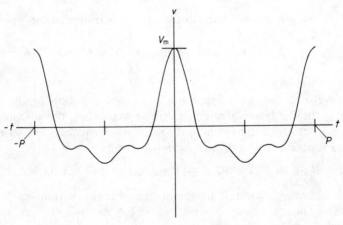

Figure 6.7

The two half-cycles of an even function waveform are related, for $t = 0 - \frac{1}{2}P$, by the equation

$$v(P - t) = v(t).$$

The even function in figure 6.7 is given by

$$v = 100 \cos \omega t + 50 \cos 2\omega t + 33 \cos 3\omega t.$$

Unidirectional impulse waveform
The waveform shown in figure 6.8 is an even function produced by summing the fundamental and consecutive harmonics all with equal amplitudes. The series for the even function impulse waveform is given by

$$v = V \cos \omega t + V \cos 2\omega t + V \cos 3\omega t + \dots V \cos n\omega t \dots$$

and theoretically requires an infinite series to produce an infinite impulse for zero time. The waveform shown in figure 6.8 contains the first twenty harmonics, with $V = 1$.

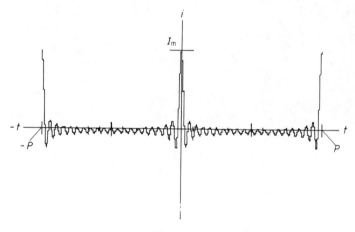

Figure 6.8

Fundamental and even harmonics only
The waveforms in figures 6.9–6.11 are produced by adding the fundamental and second harmonic with different phase relationships. In general the two half-cycles for a wave with even harmonics are not similar in shape except for the special examples of the odd and even functions described above.

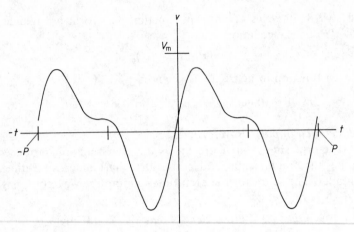

Figure 6.9

For figure 6.9

$$v = 100 \sin \omega t + 50 \sin(2\omega t + \tfrac{1}{6}\pi).$$

For figure 6.10

$$v = 100 \sin \omega t + 50 \sin 2\omega t.$$

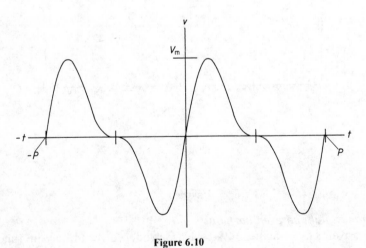

Figure 6.10

For figure 6.11,

$$v + 100 \cos \omega t + 50 \cos 2\omega t.$$

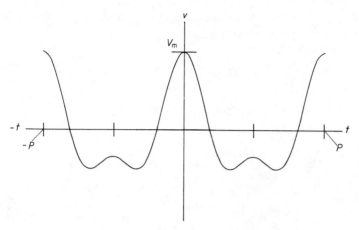

Figure 6.11

Fundamental and odd harmonics only
Waveforms with odd harmonics only are illustrated in figures 6.12–6.14. These waveforms are produced by combinations of the fundamental and third harmonic. In all cases waves with the fundamental and odd harmonics only have symmetrical half-cycles inverted with the same time sequence. This type of symmetry is described by the expression

$$v(t) = -v(t + \tfrac{1}{2}P).$$

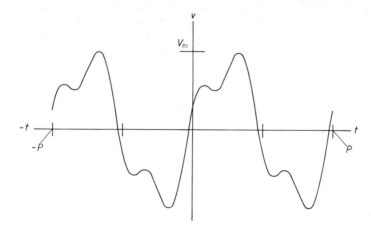

Figure 6.12

Furthermore, where the components are all sine or all cosine waves the quarter-cycles are symmetrical.

For figure 6.12,

$$v = 100 \sin \omega t + 33 \sin(3\omega t + \tfrac{1}{3}\pi).$$

For figure 6.13,

$$v = 100 \sin \omega t + 33 \sin 3\omega t.$$

For figure 6.14,

$$v = 100 \cos \omega t + 33 \cos 3\omega t.$$

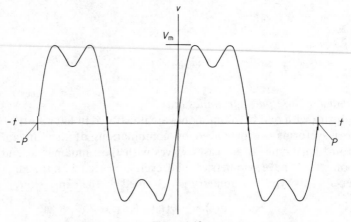

Figure 6.13

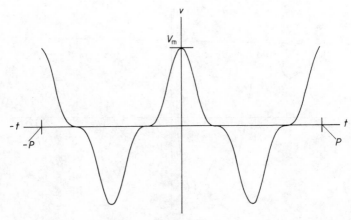

Figure 6.14

⟩ 6.3   Impulse, square and triangular waveforms

Three particular waveforms, with quarter-wave symmetry, can be synthesized by adding to the fundamental odd harmonics for which the amplitude has a simple relationship to the harmonic frequency.

These waveforms are

(a)   the positive and negative impulse waveform, which has harmonics with equal amplitudes;

(b)   the square wave, for which the harmonic amplitudes are inversely proportional to the harmonic frequency; and

(c)   the triangular waveform, for which the harmonic amplitudes are inversely proportional to the square of the harmonic frequency.

The theoretical limit for each of the waveforms requires an infinite series for its synthesis. The examples shown have a fundamental amplitude of unity and include odd harmonics up to the twenty-first.

6.3.1   Positive and negative impulse waveform synthesis

If a series of odd harmonic cosine waves of equal amplitude are added, then an impulse waveform is produced as shown in figure 6.15, with one positive and one negative impulse per cycle. The theoretical limit of each impulse is infinite magnitude for zero time. The positive and negative impulse waveform can also be synthesized as an odd function (see practice problem 1).

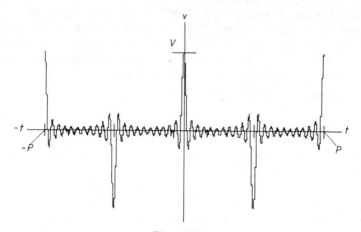

Figure 6.15

6.3.2 Square-wave synthesis

The square wave shown in figure 6.16 is synthesized by summing odd harmonic sine waves with an amplitude inversely proportional to the harmonic frequency. For the theoretical limit of the square wave the change between the constant levels occurs in zero time. The square wave can also be synthesized as an even function (see practice problem 1).

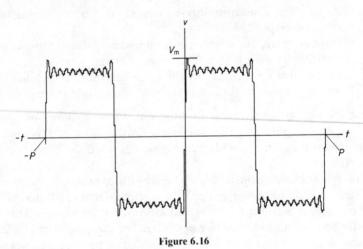

Figure 6.16

6.3.3 Triangular waveform synthesis

Figure 6.17 shows a triangular waveform produced by summing odd harmonic negative cosines, with an amplitude inversely proportional to the square of the frequency. Since the harmonic amplitudes rapidly decrease with frequency the synthesis of the triangular wave shown in figure 6.17 is the most representative of the theoretical limits for the three waveforms. The triangular waveform can also be synthesized as an odd function (see practice problem 1).

6.3.4 Application to circuit elements

The impulse, square and triangular waveforms have a differential relationship which can be illustrated in either the synthesized or harmonic component forms.

The impulse waveform, having two infinite values of opposite sign separated by a zero level, is the differential of the square wave. The square wave, which has two constant values of opposite sign, is the differential of the triangular waveform. Considered in harmonic form,

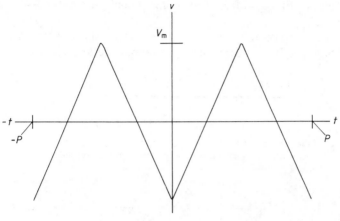

Figure 6.17

the examples of the impulse and square waveforms given consist of cosine or sine components, which are the differentials of the sine or cosine harmonics in the square and triangular waves, respectively. When impulse, square or triangular waveforms of voltage and current are applied to inductance and capacitance, the differential voltage–current relationship for the circuit elements produces the following results.

The voltage associated with a square wave of current in an inductance is an impulse wave and, with a triangular waveform of current, a square wave.

For capacitance, the current waves associated with square and triangular waveforms of voltage are respectively impulsive and square.

The responses produced by the circuit elements can alternatively be analysed in harmonic terms.

Inductive reactance is proportional to frequency and the voltage leads the current by $90°$. A triangular current wave, synthesized by negative cosines with a magnitude inversely proportional to the square of frequency, will therefore have an associated voltage drop consisting of a series of sine waves with a magnitude inversely proportional to frequency. This series produces a square wave. Similar reasoning can be used for the square wave of current and the voltage impulse.

The reactance of capacitance is inversely proportional to frequency and the current leads the voltage by $90°$. Relationships therefore exist between the three waveforms, similar to those for inductance but with voltage and current interchanged. The alternative approaches using the techniques of time and frequency-domain analysis are discussed further in §6.6.

The above responses for the circuit elements and the harmonic contents of the three waveforms are summarized in table 6.1.

Table 6.1 (*a*)

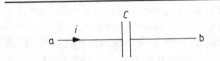

Current waveform		Voltage waveform	
Type	Harmonic amplitude	Type	Harmonic amplitude
Triangular	K_t/n^2	Square	K_s/n
Square	K_s/n	Impulse	K_i

K_t, K_s and K_i are constants
n is the *odd* harmonic number (1, 3, 5, etc)

(*b*)

a ——▶—— C —— b

Voltage waveform		Current waveform	
Type	Harmonic amplitude	Type	Harmonic amplitude
Triangular	K_t/n^2	Square	K_s/n
Square	K_s/n	Impulse	K_i

K_t, K_s and K_i are constants
n is the *odd* harmonic number (1, 3, 5, etc)

The relationships between square and triangular waves for inductance and capacitance are shown in figure 1.20. For these particular examples the triangular waves have an average value which is not zero. This is due to the time datum for the square wave coinciding with zero stored energy in the circuit elements. As indicated above, the average level does not affect the harmonic content.

The differentiating and integrating properties of circuits, additionally considered in §5.9.1, are not restricted to square and triangular waveforms. They are however, particularly apparent in these instances because the waves change in shape from one specific form to another.

6.3.5 Change in harmonic content

Any frequency-sensitive circuit will respond differently at each harmonic frequency of a complex waveform. The consequence of this will be that the harmonic content of associated voltage and current waveforms will be different. This property is particularly marked in the examples given above where the pure elements of inductance and capacitance completely change the shape of the waveforms involved.

In general, because the impedance of inductive circuits increases with frequency the percentage magnitude of each harmonic in the current waveform will be reduced relative to the value for the corresponding voltage harmonic. For capacitive circuits the reverse is true with the percentage magnitude of the current harmonics increasing as the impedance decreases with frequency.

Complex waves of current and voltage in purely resistive circuits will have the same harmonic content because in this instance the circuit properties are not frequency dependent.

⟩ 6.4 Complex wave power

The instantaneous power absorbed by a circuit represented in block form in figure 6.18 is given by

$$p = v_{ab}i.$$

If v_{ab} and i are complex waves then in general they are given by the series expressions

$$v_{ab} = V_0 + V_{1s} \sin \omega t + V_{2s} \sin 2\omega t + V_{3s} \sin 3\omega t + \dots$$
$$+ V_{1c} \cos \omega t + V_{2c} \cos 2\omega t + V_{3c} \cos 3\omega t + \dots$$

and

$$i = I_0 + I_{1s} \sin \omega t + I_{2s} \sin 2\omega t + I_{3s} \sin 3\omega t + \dots$$
$$+ I_{1c} \cos \omega t + I_{2c} \cos 2\omega t + I_{3c} \cos 3\omega t + \dots.$$

The instantaneous power expression is produced by the product of the two series and is extensive. However, the function has a regular pattern which assists the analysis.

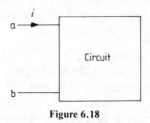

Figure 6.18

There are six types of general term in the power expression, namely,

sine terms $\qquad V_0 I_{ns} \sin(n\omega t)$, etc,
cosine terms $\qquad V_0 I_{nc} \cos(n\omega t)$, etc,
sine squared terms $\qquad V_{ns} I_{ns} \sin^2(n\omega t)$, etc,
cosine squared terms $\qquad V_{nc} I_{nc} \cos^2(n\omega t)$, etc,

sine and cosine products with the same harmonic frequencies

$$V_{ns} I_{nc} \sin(n\omega t) \cos(n\omega t), \text{ etc,}$$

and sine and cosine products with different harmonic frequencies

$$V_{ns} I_{mc} \sin(n\omega t) \cos(m\omega t), \text{ etc,}$$

where n and m are different harmonics.

By using this pattern of terms and the sum and difference relationships given in Appendix A6.4, the expression for the average power can be considerably simplified, since only the sine and cosine squared terms contribute.

The constant values V_0 and I_0, when present, produce a contribution to the average power of $V_0 I_0$, since the sine and cosine terms which they also produce have zero average values. The expression for the average power, P_{av}, is therefore

$$P_{av} = V_0 I_0 + \tfrac{1}{2} V_{1s} I_{1s} + \tfrac{1}{2} V_{2s} I_{2s} + \tfrac{1}{2} V_{3s} I_{3s} \ldots$$
$$+ \tfrac{1}{2} V_{1c} I_{1c} + \tfrac{1}{2} V_{2c} I_{2c} + \tfrac{1}{2} V_{3c} I_{3c} \ldots.$$

Reference to the analysis in Appendix A4.6 produces a further simplification, giving

$$P_{av} = V_0 I_0 + V_{1rms} I_{1rms} \cos(A_1 - B_1) + V_{2rms} I_{2rms} \cos(A_2 - B_2)$$
$$+ \ldots + V_{nrms} I_{nrms} \cos(A_n - B_n) + \ldots$$

where A_n and B_n are the phase angles of the nth harmonic, voltage and current waves, respectively, and $(A_n - B_n)$ is the phase difference between them.

6.4.1 RMS value of a complex wave

Similar analysis to that used for the power expression can be used to determine an equation for the RMS value of a complex wave. The mean square value of the voltage or current can be determined by replacing the current with voltage or the voltage with current in the P_{av} expression.

Thus the mean square value of voltage is given by

$$V_{ms} = V_0^2 + V_{1rms}^2 + V_{2rms}^2 + \ldots V_{nrms}^2 + \ldots$$

since the $\cos(A_n - B_n)$ terms are equal to 1 with B_n having been replaced by A_n.

The RMS value of the voltage wave is therefore given by the equation

$$V_{rms} = (V_0^2 + V_{1rms}^2 + V_{2rms}^2 + \ldots + V_{nrms}^2 + \ldots)^{1/2}.$$

⟩ 6.5 Waveform analysis

The process of waveform analysis is the reverse of synthesis and consists of determining the amplitude and phase of the component frequencies in a complex waveform.

The program associated with this section, which performs harmonic analysis, requires the value of the complex waveform to be known at all instances of time, as in the examples of the square and triangular waveforms described above. A similar process can, however, be applied to sampled waveforms. The method of analysis employs the property of the harmonic products given in Appendix A6.4 and used to determine average power in §6.4. The constant term in the Fourier series is the average value of the waveform.

The technique for determining the amplitudes of the sine and cosine terms for the remainder of the general series, given in §6.2.2, is to multiply the complex-wave function by the appropriate harmonic sine or cosine and then to integrate the result over one period of the fundamental. The integrals of the harmonic products are all zero except the one containing the product of the term by which the wave was multiplied and the harmonic in the series which is similar to it. The process of waveform analysis is summarized in block diagram form in figure 6.19.

Programs under filename 'HA' analyse:

(a) pulse waveforms with specified voltage levels, duty factor and period;
(b) triangular waveforms with specified maximum value, time advance and period; and
(c) full-wave rectified sine waves with specified maximum value, delay angle and period.

The integration used is described briefly in Appendix A6.5. The particular method determines the number of samples in relation to the

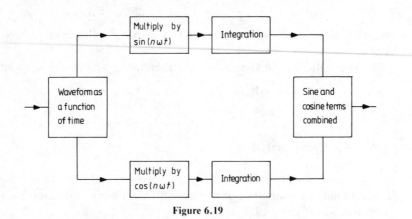

Figure 6.19

HARMONICS			HARMONICS		
No	MAGNITUDE	ANGLE	No	MAGNITUDE	ANGLE
F	1.27	0	26	0	0
2	0	0	27	4.71E-2	0
3	0.424	0	28	0	0
4	0	0	29	4.39E-2	0
5	0.255	0	30	0	0
6	0	0	31	4.1E-2	0
7	0.182	0	32	0	0
8	0	0	33	3.86E-2	0
9	0.141	0	34	0	0
10	0	0	35	3.63E-2	0
11	0.116	0	36	0	0
12	0	0	37	3.44E-2	0
13	9.79E-2	0	38	0	0
14	0	0	39	3.26E-2	0
15	8.48E-2	0	40	0	0
16	0	0	41	3.1E-2	0
17	7.49E-2	0	42	0	0
18	0	0	43	2.96E-2	0
19	6.7E-2	0	44	0	0
20	0	0	45	2.82E-2	0
21	6.06E-2	0	46	0	0
22	0	0	47	2.7E-2	0
23	5.53E-2	0	48	0	0
24	0	0	49	2.59E-2	0
25	5.09E-2	0	50	0	0

Average value=0
Fundamental F=50 hertz

Figure 6.20

required accuracy. The numerical method is itself a sampling process and is no different in principle therefore to the wave data being in sampled form. Sampled waves can only be analysed into a limited range of harmonics. The highest harmonic frequency that discrete analysis can produce is related to the sampling frequency.

Sources of more detailed information on waveform analysis are suggested in the complementary and further reading list.

HARMONICS			HARMONICS		
No	AMPLITUDE	ANGLE	No	AMPLITUDE	ANGLE
F	0.811	-90	26	0	0
2	0	0	27	1.11E-3	-90
3	9.01E-2	-90	28	0	0
4	0	0	29	9.64E-4	-90
5	3.24E-2	-90	30	0	0
6	0	0	31	8.44E-4	-90
7	1.65E-2	-90	32	0	0
8	0	0	33	7.45E-4	-90
9	1E-2	-90	34	0	0
10	0	0	35	6.62E-4	-90
11	6.7E-3	-90	36	0	0
12	0	0	37	5.93E-4	-90
13	4.8E-3	-90	38	0	0
14	0	0	39	5.34E-4	-90
15	3.6E-3	-90	40	0	0
16	0	0	41	4.83E-4	-90
17	2.81E-3	-90	42	0	0
18	0	0	43	4.39E-4	-90
19	2.25E-3	-90	44	0	0
20	0	0	45	4.01E-4	-90
21	1.84E-3	-90	46	0	0
22	0	0	47	3.68E-4	-90
23	1.53E-3	-90	48	0	0
24	0	0	49	3.38E-4	-90
25	1.3E-3	-90	50	0	0

Average value=0
Fundamental F=50 hertz

Figure 6.21

The computer printouts given in figures 6.20 and 6.21 are for analyses of square and triangular waveforms with maximum values of 1 V. The syntheses shown in figures 6.16 and 6.17 are for similar waveforms where in each case the fundamental amplitude is 1 V, i.e. for a square wave of value $\pm 1/1.27$ and a triangular wave with a maximum value of $1/0.811$.

⟩ 6.6   Frequency-domain analysis of complex waveforms

Frequency-domain analysis can be used to determine steady-state time-domain responses in circuits with complex waveforms. This is achieved by treating each harmonic as a separate source and then using the principle of superposition to determine the complete response. Having determined the response for each harmonic component of the complex wave, the steady-state time-domain waveform can be synthesized.

By using this technique the single-frequency phasor analysis developed in Chapter 4 can be applied to circuits with non-sinusoidal voltage and current sources and used as an alternative to differential equation methods for producing time-domain responses.

The two methods are demonstrated and compared as follows.

(1) Figure 6.22 is a printout from program filename 'PLW' showing the circuit parameters and response of a series $R-L$ circuit to a square wave of frequency 50 Hz and voltage levels of ± 1 V.

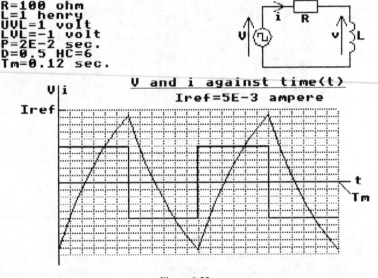

```
R=100 ohm
L=1 henry
UVL=1 volt
LVL=-1 volt
P=2E-2 sec.
D=0.5 HC=6
Tm=0.12 sec.
```

Figure 6.22

(2) Figure 6.23 is a printout from program filename 'FRP' which gives the circuit impedance at the fundamental and harmonic frequencies, up to the fifty-first.

(3) Figure 6.24 is the printout from program filename 'PHC' giving the harmonic currents. These are produced by dividing the voltage components of the square-wave analysis given in figure 6.20 by the impedance at the appropriate harmonic frequency.

(4) Figure 6.25 shows the synthesized waveform including harmonics up to the fifteenth. These are phasors P3, P6, P9, P12, P15, P18, P21 and P24 in figure 6.24.

Note that the time datum for the steady-state response produced by program filename 'SYNTH' is arbitrary and does not correspond to time

FREQUENCY	PHASOR Magnitude	deg.	FREQUENCY	PHASOR Magnitude	deg.	FREQUENCY	PHASOR Magnitude	deg.
50	329.7	72.34	900	5656	88.99	1750	1.1E4	89.48
100	636.2	80.96	950	5970	89.04	1800	1.131E4	89.49
150	947.8	83.94	1000	6284	89.09	1850	1.162E4	89.51
200	1261	85.45	1050	6598	89.13	1900	1.194E4	89.52
250	1574	86.36	1100	6912	89.17	1950	1.225E4	89.53
300	1888	86.96	1150	7226	89.21	2000	1.257E4	89.54
350	2201	87.40	1200	7540	89.24	2050	1.288E4	89.56
400	2515	87.72	1250	7855	89.27	2100	1.32E4	89.57
450	2829	87.97	1300	8169	89.30	2150	1.351E4	89.58
500	3143	88.18	1350	8483	89.32	2200	1.382E4	89.59
550	3457	88.34	1400	8797	89.35	2250	1.414E4	89.59
600	3771	88.48	1450	9111	89.37	2300	1.445E4	89.60
650	4085	88.60	1500	9425	89.39	2350	1.477E4	89.61
700	4399	88.70	1550	9739	89.41	2400	1.508E4	89.62
750	4713	88.78	1600	1.005E4	89.43	2450	1.539E4	89.63
800	5028	88.86	1650	1.037E4	89.45	2500	1.571E4	89.64
850	5342	88.93	1700	1.068E4	89.46	2550	1.602E4	89.64

Figure 6.23

CALCULATION	PHASOR No.	Magnitude	deg.	No.	Magnitude	deg.	No.	Magnitude	deg.
P3=P1/P2	P0	1	0.000	P20	4085	88.60			
P6=P4/P5	P1	1.27	0.000	P21	2.397E-5	-88.60			
P9=P7/P8	P2	329.7	72.34	P22	8.48E-2	0.000			
P12=P10/P11	P3	3.852E-3	-72.34	P23	4713	88.78			
P15=P13/P14	P4	0.424	0.000	P24	1.799E-5	-88.78			
P18=P16/P17	P5	947.8	83.94						
P21=P19/P20	P6	4.474E-4	-83.94						
P24=P22/P23	P7	0.255	0.000						
	P8	1574	86.36						
	P9	1.62E-4	-86.36						
	P10	0.102	0.000						
	P11	2201	87.40						
	P12	8.269E-5	-87.40						
	P13	0.141	0.000						
	P14	2829	87.97						
	P15	4.984E-5	-87.97						
	P16	0.116	0.000						
	P17	3457	88.34						
	P18	3.356E-5	-88.34						
	P19	9.79E-2	0.000						

Figure 6.24

$t = 0$ for the initial response produced by program filename 'PLW'. This is a general point which should always be considered when viewing steady-state waveforms.

WAVEFORM SYNTHESIS

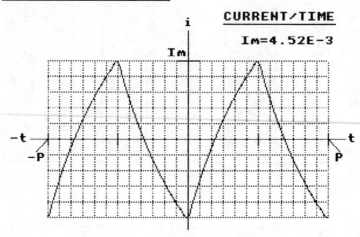

Figure 6.25

6.6.1 Harmonic resonance

A complex wave may cause a circuit, or system, to resonate at the fundamental frequency or a higher harmonic. For a supply system this effect could produce unexpectedly high harmonic currents and/or voltages. Alternatively, a high-Q-factor circuit can be designed to deliberately select a specific harmonic from a complex waveform.

Figure 6.26 illustrates the effect of supplying a series $R-L-C$ circuit with a square wave, the frequency of which is one third that of the resonant frequency of the circuit. The Q factor of the circuit is approximately 100 and the current response is almost entirely third harmonic.

The current waveform, obtained using program filename 'PLW', shows the first six cycles of the response with the third harmonic increasing from zero towards a steady state. The steady-state response can be obtained by frequency-domain analysis as previously described and shown in figures 6.27 and 6.28.

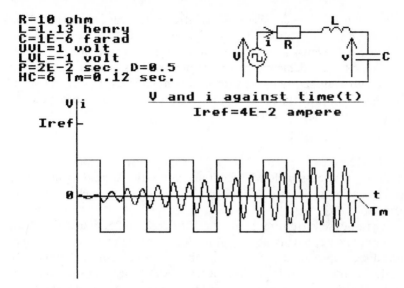

Figure 6.26

FREQUENCY	PHASOR Magnitude	deg.	FREQUENCY	PHASOR Magnitude	deg.	FREQUENCY	PHASOR Magnitude	deg.
50	2829	-89.88	220	832.8	89.31	390	2351	89.76
60	2228	-89.74	230	935	89.39	400	2432	89.76
70	1779	-89.68	240	1035	89.45	410	2512	89.77
80	1424	-89.68	250	1132	89.49	420	2592	89.78
90	1132	-89.49	260	1227	89.53	430	2672	89.79
100	884.2	-89.35	270	1320	89.57	440	2751	89.79
110	668.8	-89.14	280	1412	89.59	450	2829	89.80
120	477.6	-88.88	290	1503	89.62	460	2908	89.80
130	304.9	-88.12	300	1592	89.64	470	2986	89.81
140	146.9	-86.18	310	1679	89.66	480	3064	89.81
150	10	0.049	320	1766	89.68	490	3141	89.82
160	137.4	85.83	330	1852	89.69	500	3219	89.82
170	266.5	87.05	340	1937	89.70	510	3295	89.83
180	389.2	88.53	350	2021	89.72	520	3372	89.83
190	586.4	88.87	360	2104	89.73	530	3449	89.83
200	619	89.07	370	2187	89.74	540	3525	89.84
210	727.6	89.21	380	2269	89.75	550	3601	89.84

IMPEDANCE VARIATION WITH FREQUENCY

MAGNITUDE/FREQUENCY

ANGLE/FREQUENCY

Figure 6.27

CALCULATION	PHASOR No.	Magnitude	deg.	No.	Magnitude	deg.	No.	Magnitude	deg.
P3=P1/P2	P0	1	0.000						
P6=P4/P5	P1	1.27	0.000						
P9=P7/P8	P2	2829	-89.00						
P12=P10/P11	P3	4.489E-4	89.00						
	P4	0.424	0.000						
	P5	10	0.049						
	P6	4.24E-2	-0.049						
	P7	0.255	0.000						
	P8	1132	89.49						
	P9	2.253E-4	-89.49						
	P10	0.182	0.000						
	P11	2021	89.72						
	P12	9.005E-5	-89.72						

WAVEFORM SYNTHESIS

CURRENT/TIME

$Im=4.31E-2$

Figure 6.28

Figure 6.27 shows the impedance/frequency characteristic of the circuit. These are printouts from program filename 'FRP'.

Figure 6.28 shows a printout from program filename 'PHC', used to determine the fundamental and harmonic currents and the synthesis of the steady-state current waveform which includes the fundamental together with the third, fifth and seventh harmonics, produced by program filename 'SYNTH'.

6.6.2 Accuracy

Accuracy is discussed elsewhere in the text but the techniques illustrated in this section serve as a further focus for a subject which needs to be stressed.

The response waveforms shown in figures 6.22, 6.25, 6.26 and 6.28 are all approximate, having been computed by a range of techniques which introduce approximations at a number of stages in the process of producing the results. However, the accuracy of the results produced by such techniques can be continuously increased if it is necessary in relation to the extra computing required. For example, waveform display comparable with oscilloscope accuracy is not exacting, whereas the need for more precisely quantified results may warrant extra computation.

In specific terms, the numerical techniques used for solving differential equation models of circuits for time-domain analysis are subject to local errors at each step and to accumulated global errors. The latter may become particularly large if the process involves many computations with local errors which would otherwise be acceptable. Each of these types of error can be reduced by decreasing the integration interval.

The waveform synthesis technique for producing steady-state time-domain responses uses approximate wave analysis based on a numerical technique, which is subject to similar errors to those described above and in addition includes a limited range of harmonics. Again the accuracy can be continuously increased by reducing integration intervals and increasing the range of harmonics included in the analysis. The accuracy of the phasor processes used, as with all other computations, depends on the inherent accuracy of the computer and any truncation errors built into the programs.

For a more detailed analysis of errors consult the sources of information listed in the complementary and further reading list.

⟩ 6.7 Introduction to non-linear circuit analysis

The models of the circuit elements considered to this point in the book have been linear, as described in §1.4.5. For many applications these models are appropriate for real systems. There are, however, many applications where the circuit components are non-linear.

Non-linear characteristics can be exploited by circuit designers to produce required results but in other instances may cause unwanted side effects. Some of the properties and analysis techniques associated with non-linear circuits are illustrated below, using the diode as an example.

6.7.1 Diode characteristics

The current–voltage (i–v) characteristic for a diode is shown in figure 6.29 with three superimposed linear approximations. The approximations are progressively more accurate and can be described as follows.

(a) This is the ideal switch characteristic with the two straight line portions of the characteristic coinciding with the voltage axis for negative values of voltage and with the current axis for positive voltages. Hence no current flows in the reverse direction and there is no voltage drop in the forward conduction direction. This

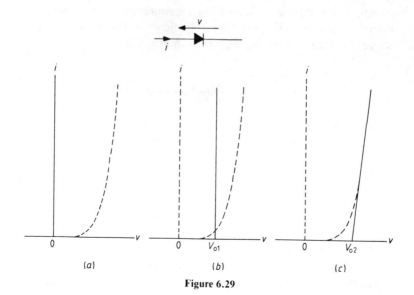

Figure 6.29

characteristic can be described by the relationships $i = 0$ for $v < 0$, $v = 0$ for $i > 0$.

(b) Actual diodes have a small reverse current and the forward current remains low until the cut-in voltage is reached. This is quoted as being about 0.6 V for silicon diodes. The second linear characteristic therefore has an offset along the voltage axis to take account of the cut-in voltage. The characteristic is described by the expressions $i = 0$ for $v < V_{01}$, $v = V_{01}$ for $i > 0$.

(c) The third linear approximation of the diode characteristic additionally allows for the forward slope resistance which is due to the increased voltage drop across the diode as the forward current increases. The expressions which describe this characteristic are $i = 0$ for $v < V_{02}$, $v = V_{02} + R_s i$ for $v > V_{02}$, where $1/R_s$ is the slope of the characteristic for voltages $> V_{02}$.

The diode has other non-linear operating regions than those shown, which extend the i–v characteristic in both directions. These regions could be similarly approximated by linear sections. The number of linear sections used to represent the non-linear characteristic depends on the required accuracy.

6.7.2 Diode–resistance circuit

The method of analysing a circuit using an approximate piece-wise linear characteristic model, for a non-linear element, is demonstrated by reference to the diode–resistance circuit in figure 6.30.

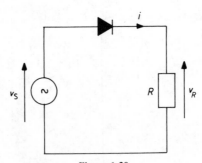

Figure 6.30

If V_S is the RMS value of the source voltage and v_S and v_R the instantaneous values of the source and resistance voltages, respectively, then the following analysis applies:

$$V_m = \sqrt{2}\, V_S$$
$$v_S = V_m \sin \omega t.$$

V_0 is the offset voltage for the linearized characteristic. If $v_S < V_0$ then

$$i = 0.$$

If $v_S > V_0$ then

$$v_S - v_R = V_0 + R_s i$$
$$i = (v_S - V_0)/(R_s + R)$$
$$v_R = (v_S - V_0)R/(R_s + R).$$

Program filename 'DIODE' produces waveforms for v_S, i and v_R against time. The parameters of the circuit can be varied as described in the printout of the preamble to the program, shown in figure 6.31.

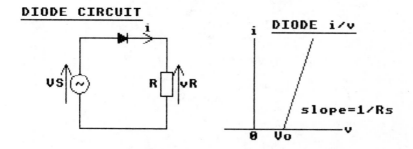

DIODE CIRCUIT

DIODE i/v

slope=1/Rs

This program draws the voltage/time and current/time graphs, for vR and i.

The diode operates with the characteristic shown.

The parameters US(volt rms), Uo(volt), Rs(ohm) and R(ohm), are all variable.

Figure 6.31

Small values of input voltage

For small input voltages the offset voltage is significant and figure 6.32 shows the effect of the three levels of linear approximation in figure 6.29.

For the ideal switch type characteristic, v_R is coincident with v_S for the positive half cycle of the source voltage. All three characteristics produce zero values of i and v_R, for the negative half-cycle of the source voltage.

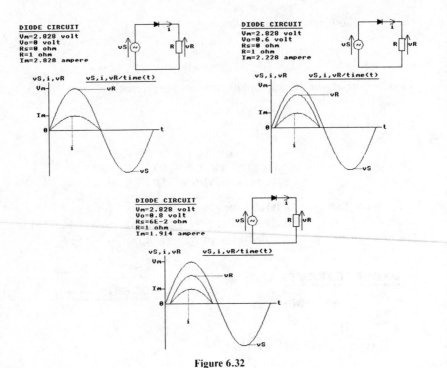

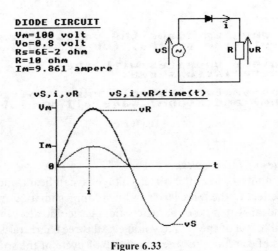

Figure 6.32

Figure 6.33

Large values of input voltage
For large values of input voltage the offset voltage will be negligible in relative terms. If R_s can also be ignored in relation to R, then all three models produce the same result. Figure 6.33 illustrates this by showing that, within the resolution of the graph, v_S and v_R are coincident for the positive half cycle of the source voltage.

6.7.3 Diode bridge circuit
The diode bridge circuit shown in figure 6.34 produces a undirectional current flow in the resistance, R, during both half cycles of the source voltage. For each half cycle two of the diodes on opposite sides of the bridge conduct; the other two being reversed biased are non-conducting.

The circuit produces a full-wave rectified sine wave of voltage across resistance R, as shown. If the diodes are assumed to be ideal switches then the maximum value of the rectified sine wave is equal to the source voltage amplitude. It should be noted that the fundamental frequency of the full-wave rectified sine wave is twice that of the waveform rectified.

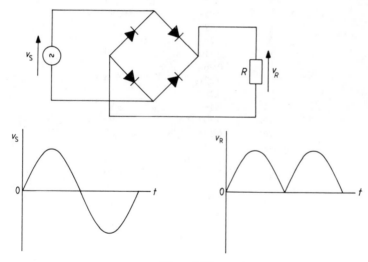

Figure 6.34

6.7.4 Harmonic distortion
Table 6.2 shows the harmonic components in a full-wave rectified sine wave, with a delay angle of zero and a maximum value of 100 V. The harmonic analysis shows a general feature of non-linear circuits, namely

the generation of harmonics not present in the source waveform. All frequency-sensitive circuits change the ratio of the voltage and current wave harmonics but non-linear systems create extra harmonics; this is referred to as harmonic distortion.

Table 6.2 The harmonic content shown was determined using program filename 'HA'. The analysis of a full-wave rectified sine wave can be quoted in terms of a fundamental frequency equal to that of the unrectified wave, in which case the component frequencies in table 6.2 correspond to even harmonics with no fundamental, i.e. the frequency of the unrectified sine wave is 50 Hz.

Harmonics		
No	Amplitude	Angle
F	42.4	− 90
2	8.49	− 90
3	3.64	− 90
4	2.02	− 90
5	1.29	− 90
6	0.89	− 90
7	0.653	− 90
8	0.499	− 90
9	0.394	− 90
10	0.319	− 90
11	0.264	− 90
12	0.221	− 90
13	0.189	− 90
14	0.163	− 90
15	0.142	− 90
16	0.124	− 90
17	0.118	− 90
18	9.83×10^{-2}	− 90
19	8.83×10^{-2}	− 90
20	7.97×10^{-2}	− 90
21	7.22×10^{-2}	− 90
22	6.58×10^{-2}	− 90
23	6.02×10^{-2}	− 90
24	5.53×10^{-2}	− 90
25	5.1×10^{-2}	− 90

Average value = 63.7
Fundamental F = 100 Hz

6.7.5 An alternative approach to non-linear analysis
An alternative approach to non-linear circuit analysis, used in simulation

programs, models non-linear elements by equations relating the device current to its terminal voltage. Equations for diodes and other semiconductor devices are typically derived from a theoretical analysis of the physical behaviour of an idealized element, possibly modified empirically for an actual device.

The current–voltage equation used for the diode in this illustration of the analysis technique is

$$i = I_s [\exp(v/V) - 1] \text{ A}$$

where i and v are the diode current and voltage drop, I_s is the saturation current and V is a temperature-dependent device parameter. The typically quoted values for I_s and V used as default values in the analysis program are

$$I_s = 10^{-14} \text{ A and } V = 25 \times 10^{-3} \text{ V at } 20\,^{\circ}\text{C}.$$

The demonstration of this analytical approach is based on the series diode–resistance circuit in figure 6.35. The current in the non-linear diode and linear resistance is governed by the models for each element. Thus, $i = I_s [\exp(v/V) - 1]$ for the diode and $i = (V_s - v)/R$ for the resistance. Since for the series circuit these currents must be equal

$$I_s [\exp(v/V) - 1] - (V_s - v)/R = 0$$

and the operating point for the circuit is determined by the value of v which satisfies the equation.

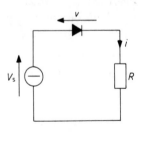

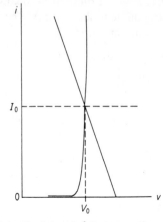

Figure 6.35 $i = [I_s \exp(v/V) - 1] = (V_s - v)R$ A, $V = 2.5 \times 10^{-2}$ V, $I_s = 10^{-14}$ A, $V_s = 0.8$ V, $R = 10^4$ Ω, $I_0 = 2.58 \times 10^{-5}$ A, $V_0 = 0.542$ V.

Program filename 'DA' solves the above equation for the circuit shown in figure 6.35 for specified values of source voltage V_s and resistance R. The program additionally allows I_s and V to be varied within prescribed limits. The program prints out the operating values of the diode current and voltage drop and then graphs the non-linear and linear characteristics for the diode and resistance, respectively. The circuit current is the value at the intersection of the two characteristics where they are both satisfied.

An example of the results from program filename 'DA' is shown in figure 6.35. A brief description of the algorithm used for the computation is given in Appendix A6.7.5.

For more information on semiconductor device models used in simulation programs, consult the suggested texts given in the complementary and further reading list.

6.7.6 Other non-linear components
Apart from semiconductor devices, another common class of non-linear components are inductors and other magnetic field devices which utilize ferromagnetic materials. Ferromagnetic materials saturate and have non-linear characteristics which, unlike those for diodes, are not generally directional. The likely consequence of such characteristics is the production of symmetrical complex waves containing odd harmonics.

Summary
This introduction to non-linear circuit analysis has used the characteristic of the diode to illustrate two basic approaches to solving circuit problems involving non-linear elements. These are as follows.

(1) The process of linearly interpolating intermediate points on a non-linear characteristic, the range of the interpolation being as small as the accuracy requires.
(2) Representing the non-linear element by a functional model which relates the device current to its terminal voltage and then solving the circuit equations produced.

For either method, the complex wave responses of non-linear circuits with repetitive waveform sources can be analysed using the techniques described in earlier sections of this chapter. An example where these techniques can be applied is the rectifier filter circuit given as practice problem 7.

> **6.8 Software development**

6.8.1 Electronic computer-aided design (ECAD)

The software tools designed for this CIT begin at the opposite end of the spectrum to the industry standard and intermediate levels of ECAD packages which are available. They are intended to solve relatively specific problems, be easy to use and leave the user in touch with how they operate. ECAD tools on the other hand are designed to solve a broad range of large-scale problems with a corresponding magnification in size and complexity of the hardware and software involved. This involves the user in a considerable investment of time in learning to use the tools effectively and regular practice to maintain that skill. The complexity of the system can leave the user remote and unaware of how the software produces the solution to a specific problem.

In this chapter the use of a group of programs together to perform a larger scale piece of analysis is demonstrated and it is easy to see how the software could be developed to make such processes automatic and more flexible. Such development would eventually lead to a full-scale ECAD package but the original objectives would be lost and existing work probably duplicated. This suggests that there is probably an optimum size and complexity for this type of CIT software.

6.8.2 Program extensions

There will probably be no definitive answer to the question of how sophisticated the software designed for the student of a subject needs to be in order to successfully underpin the use of software designed for the practising professional, while being a useful problem-solving tool in its own right. It seems appropriate therefore to build into the CIT flexibility in the form of scope for program extensions. Two program filenames 'FDSD' and 'TDSD' are therefore reserved for frequency- and time-domain software development.

The mesh and node analysis programs filenames 'MAZC' and 'NAYC' have an inbuilt facility for changing the analysis frequency and this can be extended to produce results over a frequency range. The state-variable analysis introduced in Chapter 5 can be extended to include more state variables and time-varying sources.

Each of these techniques can be made more general by the programs computing output responses which are combinations of mesh currents, node voltages or state variables, respectively.

These examples, and it is also likely to be true of other program developments, require no further text expansion of circuit theory principles and it is appropriate that they should be carried by the software alone.

〉 6.9   General conclusions

In §4.4.3 a brief comparison was made between specific computational and analytical methods and it seems appropriate to conclude this book with a broader comparison.

The development of electric circuit theory has produced a range of sophisticated and well established analytical tools to which computing methods are being added. The combination of these two approaches will lead to the synthesis of a more effective theoretical and problem-solving toolkit for engineers. The programs included in this CIT illustrate some of the ways in which this process may progress. These can be broadly categorized as: aiding the solution of equations produced by traditional techniques, employing modified models more appropriate to computer use and exploiting a particular strength of computing methods to add another dimension to analysis and simulation.

Another likely consequence of the amalgamation of the two approaches will be the erosion of any artificial boundaries between sections of theory and the channelling of perception imposed by the limitations of particular analytical tools.

The essence of computer methods, requiring an awareness of the appropriateness, limitations and accuracy of numerical methods and the interpretation of graphical information, has much in common with skills required in the practical world of experimental measurement and instrumentation. As such they therefore provide a suitable environment for the student engineer.

The right conditions would therefore seem to exist for generating a period of continuing development in circuit theory at a level where the subject has for some time been relatively static. The pressure for change will come from the ever-increasing availability and dispersal of computing power and the changes themselves from a creative response to that pressure.

⟩ 6.10   Practice problems

1   (i)   Use program filename 'SYNTH' to synthesize the following:

(1)   a sawtooth waveform as an odd function with a positive slope which passes through the origin of the voltage and time axes;

(2)   a undirectional negative impulse wave;

(3)   a square wave as an even function;

(4)   a positive and negative impulse waveform as an odd function;

(5)   a triangular waveform as an odd function;

(6)   a full-wave rectified sine wave as an even function with the maximum value on the voltage axis.

(ii)   Demonstrate that the undirectional impulse waveform is the differential of the sawtooth waveform.

2   The four waveforms shown in figure 6.36 consist of equal amplitude, sine and cosine, fundamental and second harmonic products. Identify each waveform with the appropriate product. Use program filename 'SYNTH' to confirm the sum and difference relationships given in Appendix A6.4 by synthesizing the waveforms with two harmonics.

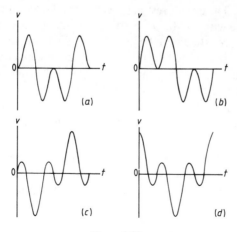

Figure 6.36

3   Two pulse-width-modulated 1 kHz supplies are used to provide a
    varying level of average voltage. One supply switches between $+24$
    and -24 V, the other between $+24$ and 0 V. Determine the duty
    factor for each waveform, required to produce an average level of
    14.4 V. Draw a line spectrum for each waveform and hence compare
    their harmonic content at this operating voltage.

4   Write a simple program loop to illustrate the RMS relationship for a
    complex wave by using the specific example of a square wave with a
    maximum value of 1 V. Perform the computation for the first one
    hundred harmonics.

5   A 1 kHz square-wave source of maximum value 1 V is applied to a
    series $R-L$ circuit where $R = 10$ Ω and $L = 1$ mH.

    (i)   Produce the time-domain steady-state current waveform for
            the circuit using program filename 'PLW'.
    (ii)   Use the method described in §6.6 to synthesize the current
            waveform using frequency-domain analysis.
    (iii)   Determine the contribution to the total power from the funda-
            mental and the third, fifth and seventh harmonics.

6   A series $R-L-C$ circuit is connected to a square-wave voltage source
    of ± 0.8 V. The circuit resonates at the fundamental frequency of the
    square wave with a Q factor of 100. Consider three circuit conditions,
    each with a resistance of 10 Ω and an inductance of 1.59, 0.159 and
    0.0159 H respectively. Determine the value of capacitance required in
    each case and the steady-state amplitude of the fundamental current.
    Plot a graph of the amplitude of the fundamental current against (i)
    time and (ii) square-wave cycle number, to demonstrate the rate at
    which the amplitude builds up from zero to its steady-state value.

7   A bridge rectifier circuit is shown in figure 6.37 where the diodes can
    be assumed to be ideal switches. The sinusoidal source voltage, V_S, is
    70.7 V RMS at 50 Hz. Synthesize the waveform for v_R with and
    without the average value present and determine the peak–peak
    voltage ripple and the attenuation produced in the fundamental of
    the rectified waveform by the $L-C$ filter circuit.

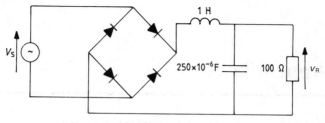

Figure 6.37

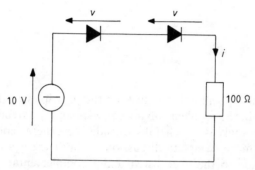

Figure 6.38

8 Determine the voltage drop v across each diode and the current i for
   the circuit shown in figure 6.38. The diodes can be assumed to be
   identical and modelled by the default parameters used in program
   filename 'DA'.

> Appendix A

> Further Text Development

> Introduction

The aim of this appendix is to illustrate the principles of the numerical methods used in the problem-solving software and to extend some of the circuit theory analysis which falls outside the general concept of this book. For a more complete discussion of both these areas reference should be made to the texts listed under complementary and further reading. The sections in this appendix are numbered to correspond with the relevant parts of the main text.

> 1.6   Mutual inductance: parallel equivalents

The equivalent inductances for the parallel connection of inductances with mutual inductance are derived in Table A1.

> 1.10   Solution of simultaneous equations

There are a number of methods for solving simultaneous equations. Cramer's rule is illustrated below for two unknown quantities. Methods suitable for a larger number of variables are given in later sections.

For simultaneous equations expressed in the form

$$V_1 = R_1 I_1 + R_2 I_2$$
$$V_2 = R_3 I_1 + R_4 I_2$$

Table A1 Inductances with mutual inductance in parallel.

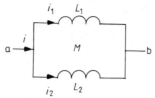

$di/dt = di_1/dt + di_2/dt$
$V_{ab} = L_1\, di_1/dt \pm M\, di_2/dt$
$V_{ab} = \pm M\, di_1/dt + L_2\, di_2/dt.$

Using the rule given below in Appendix A1.10

$$di_1/dt = \begin{vmatrix} V_{ab} & \pm M \\ V_{ab} & L_2 \end{vmatrix} \begin{vmatrix} L_1 & \pm M \\ \pm M & L_2 \end{vmatrix}^{-1} \qquad di_2/dt = \begin{vmatrix} L_1 & V_{ab} \\ \pm M & V_{ab} \end{vmatrix} \begin{vmatrix} L_1 & \pm M \\ \pm M & L_2 \end{vmatrix}^{-1}.$$

But $V_{ab} = L\, di/dt$ where L is the equivalent inductance, hence
$di/dt = (L_2 L\, di/dt \mp ML\, di/dt + L_1 L\, di/dt \mp ML\, di/dt)/(L_1 L_2 - M^2)$
$L = (L_1 L_2 - M^2)/(L_1 + L_2 \pm 2M)$

the unknown quantities can be expressed as the ratio of two determinants:

$$I_1 = \begin{vmatrix} V_1 & R_2 \\ V_2 & R_4 \end{vmatrix} \begin{vmatrix} R_1 & R_2 \\ R_3 & R_4 \end{vmatrix}^{-1} \qquad I_2 = \begin{vmatrix} R_1 & V_1 \\ R_3 & V_2 \end{vmatrix} \begin{vmatrix} R_1 & R_2 \\ R_3 & R_4 \end{vmatrix}^{-1}.$$

⟩ 2.3 Matrix multiplication and equation solution

The process of matrix multiplication is shown below.

$$\begin{bmatrix} R_1 & R_2 & R_3 \\ R_4 & R_5 & R_6 \\ R_7 & R_8 & R_9 \end{bmatrix} \begin{bmatrix} I_1 \\ I_2 \\ I_3 \end{bmatrix} = \begin{bmatrix} V_1 \\ V_2 \\ V_3 \end{bmatrix} \quad \text{means} \quad \begin{array}{l} R_1 I_1 + R_2 I_2 + R_3 I_3 = V_1 \\ R_4 I_1 + R_5 I_2 + R_6 I_3 = V_2 \\ R_7 I_1 + R_8 I_2 + R_9 I_3 = V_3 \end{array}$$

It can be seen that each row of the matrix equation corresponds to one of the set of simultaneous equations.

Gaussian elimination
A set of simultaneous equations similar to that shown above is reduced to the form shown below. This is achieved by adding an appropriately

scaled version of one equation to those remaining in the set. Once in the form shown the unknown parameters can be determined by back substitution.

$$R_1 I_1 + R_2 I_2 + R_3 I_3 = V_1$$
$$R_{10} I_2 + R_{11} I_3 = V_4$$
$$R_{12} I_3 = V_5.$$

Gauss–Seidel iteration
A set of equations similar to those in above is rearranged in the form shown below and then solved sequentially and iteratively to converge on the final values

$$I_1 = (V_1 - R_2 I_2 - R_3 I_3)/R_1$$
$$I_2 = (V_2 - R_4 I_1 - R_6 I_3)/R_5$$
$$I_3 = (V_3 - R_7 I_1 - R_8 I_2)/R_9.$$

It should be noted that all sets of equations will not necessarily produce the expected results and reference should be made to appropriate texts for more detailed presentations of these and other numerical methods.

⟩ 2.7.1 Maximum power by calculus

The expression for power transfer given in §2.7 is

$$P = V_{oc}^2 R_x [V_{oc}/(R + R_x)^2]$$

hence

$$dP/dR_x = V_{oc}^2 [(R + R_x)^{-2} - 2 R_x (R + R_x)^{-3}].$$

$dP/dR_x = 0$ when P is a maximum. This occurs when $(R + R_x) - 2R_x = 0$, therefore $R_x = R$.

⟩ 2.7.2 Maximum power by computer

The circuit parameters refer to figure 2.17. This algorithm is based on the knowledge gained from a simple analysis of the circuit that the power transferred to R_x has a maximum value between $R_x = 0$ and $R_x = $ infinity.

Thus after each increment of R_x from zero, the power is computed and compared with the previous value until the new value is less than the previous one, which means the maximum value has been passed.

The outline program given below uses a linear increase for R_x, a logarithmic one is also appropriate.

Old power = OP
New power = NP
Max power = MP

NP = 0: Rx = 0.1 (or required value)

REPEAT
  OP = NP
  NP = Rx*((Voc/(R + Rx))^2)
  IF NP > OP THEN MP = NP ELSE MP = OP
  Rx = Rx + 0.1(or required increment)
UNTIL MP = OP

PRINT MP

PRINT Rx − 0.1(or required increment).

⟩ **3.2.3 Half-cycle average and RMS values**

Half cycle average
For the half cycle of the sine wave shown in figure A1

$$v = V_m \sin(\omega t).$$

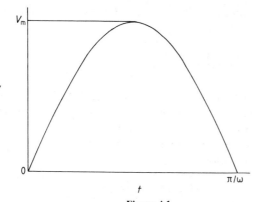

Figure A1

Hence the average value over half of one cycle of the complete sine wave is given by

$$V_{av} = \left(\int_{t=0}^{t=\pi/\omega} V_m \sin(\omega t) \, dt \right) (\pi/\omega)^{-1}$$

$$V_{av} = - \left|_{t=0}^{t=\pi/\omega} [(V_m/\omega) \cos(\omega t)] (\pi/\omega)^{-1}.$$

Since $\cos(\pi) = -1$ and $\cos(0) = 1$

$$V_{av} = (2/\pi) V_m.$$

Hence the half-cycle average value $= 2/\pi$ times the maximum value.

Note that if the negative half cycle is removed and the average taken over one cycle then the average value is half of the above figure.

RMS value

If the instantaneous value of a sine wave is given by

$$v = V_m \sin(\omega t)$$

then

$$v^2 = V_m^2 \sin^2(\omega t) = V_m^2 [1 - \cos(2\omega t)]/2.$$

The average value of $\cos(2\omega t)$ over one cycle of $\sin(\omega t)$ is equal to zero.

Therefore

$$\text{mean of } v^2 = V_m^2/2$$

and

$$V_{rms} = \sqrt{(V_m^2/2)} = V_m/\sqrt{2}.$$

Hence for a sine wave, the RMS value = the maximum value $/\sqrt{2}$.

⟩ **4.5.4 Matrix partitioning**

Programs with filenames 'MAZ', 'MAZC', 'NAY' and 'NAYC' use partitioned matrices with reference and quadrature components to perform mesh and node analysis with phasors.

The principle of the method can be illustrated by reference to the following matrix equation:

$$\begin{bmatrix} V_1 \\ V_2 \end{bmatrix} = \begin{bmatrix} Z_{11} & Z_{12} \\ Z_{21} & Z_{22} \end{bmatrix} \begin{bmatrix} I_1 \\ I_2 \end{bmatrix}$$

$V_1 = V_{1rdeg0} + V_{1qdeg90}$ and $I_1 = I_{1rdeg0} + I_{1qdeg90}$

$V_2 = V_{2rdeg0} + V_{2qdeg90}$ and $I_2 = I_{2rdeg0} + I_{2rdeg90}$

$Z_{11} = R_{11deg0} + X_{11deg90}$ and $Z_{12} = R_{12deg0} + X_{12deg90}$

$Z_{21} = R_{21deg0} + X_{21deg90}$ and $Z_{22} = R_{22deg0} + X_{22deg90}.$

Expanding the top row of the matrix gives

$$V_{1rdeg0} + V_{1qdeg90} = R_{11}I_{1rdeg0} + X_{11}I_{1rdeg90} + R_{11}I_{1qdeg90} + X_{11}I_{1qdeg180}$$

$$+ R_{12}I_{2rdeg0} + X_{12}I_{2rdeg90} + R_{12}I_{2qdeg90} + X_{12}I_{2qdeg180}.$$

The two components can be equated separately and the angles ignored since the fact that they are components allows for the angles. Components with angles of 180° are negative reference ones.

Hence

$$V_{1r} = R_{11}I_{1r} - X_{11}I_{1q} + R_{12}I_{2r} - X_{12}I_{2q}$$

and

$$V_{1q} = X_{11}I_{1r} + R_{11}I_{1q} + X_{12}I_{2r} + R_{12}I_{2q}.$$

The second row of the matrix can be similarly expanded and the four equations expressed in matrix form as follows:

$$\begin{bmatrix} V_{1r} \\ V_{2r} \\ V_{1q} \\ V_{2q} \end{bmatrix} = \begin{bmatrix} R_{11} & R_{12} & -X_{11} & -X_{12} \\ R_{21} & R_{22} & -X_{21} & -X_{22} \\ X_{11} & X_{12} & R_{11} & R_{12} \\ X_{21} & X_{22} & R_{21} & R_{22} \end{bmatrix} \begin{bmatrix} I_{1r} \\ I_{2r} \\ I_{1q} \\ I_{2q} \end{bmatrix}$$

Using this method the components can be determined by normal algebraic methods, as in Appendix A2.3, and then recombined to produce phasor results.

⟩ 4.6 General power analysis

The analysis of power in §4.6.1 was produced with the current as the reference phase. If both voltage and current waves have phase angles, a similar method of analysis can be used which leads to a more general result.

The instantaneous values of voltage drop and current at the terminals of a single-phase circuit are given by

$$v = V \sin(\omega t + A)$$
$$i = I \sin(\omega t + B).$$

Both voltage and current can be expressed in component terms, giving

$$v = V_s \sin \omega t + V_c \cos \omega t$$
$$i = I_s \sin \omega t + I_c \cos \omega t$$

where
$$V_s = V \cos A \qquad\qquad V_c = V \sin A$$
$$I_s = I \cos B \qquad\qquad I_c = I \sin B.$$

The expression for the instantaneous power p is therefore

$$p = V_s I_s \sin^2(\omega t) + V_c I_s \cos(\omega t) \sin(\omega t) + V_s I_c \sin(\omega t) \cos(\omega t) + V_c I_c \cos^2(\omega t).$$

The sum and difference relationships given in Appendix A6.4 can now be used to determine the average power, P_{av}, which is given by

$$P_{av} = V_s I_s / 2 + V_c I_c / 2 = VI(\cos A \cos B + \sin A \sin B)/2.$$

Hence,
$$P_{av} = VI \cos(A - B)/2 = V_{rms} I_{rms} \cos(A - B).$$

The power factor in the general case is therefore the cosine of the phase difference between the voltage drop and the current.

⟩ **5.3   Step input to $R-L$ circuit**

For the circuit shown in figure A2 a step voltage of V V is applied at time $t = 0$. The source voltage is then reduced to zero at time $t = $ infinity.

For time $t = 0$ to infinity, the circuit analysis is given by

$$V = Ri + L \, di/dt$$

$$di/dt = (R/L)[(V/R) - i]$$

$$\int \{1/[(V/R) - i]\} di = (R/L) \int dt$$

$$-\ln[(V/R) - i] = (R/L)t + A$$

$$(V/R) - i = B \exp[-(R/L)t].$$

At $t = 0_-$ $i = 0$, hence at $t = 0_+$ $i = 0$ and $B = V/R$, since the current cannot change instantaneously in inductive circuits.

Thus

$$i = (V/R)\{1 - \exp[-(R/L)t]\} = (V/R)[1 - \exp(-t/T)] \text{ A}$$

where $T = L/R$ s.

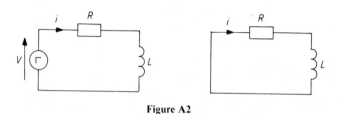

Figure A2

From the instant the voltage source is reduced to zero, the circuit analysis is given by

$$0 = Ri + L \; di/dt$$

$$di/dt = -(R/L)i$$

$$\int (1/i) \; di = -(R/L) \int dt$$

$$\ln(i) = -(R/L)t + C$$

$$i = D \exp[-(R/L)t].$$

If at the instant the voltage source is reduced to zero a new datum for time is chosen then at $t = 0_-$ $i = V/R$ and at $t = 0_+$ $i = V/R$ and $D = V/R$.

Hence

$$i = (V/R)\exp[-(R/L)t] = (V/R)\exp(-t/T) \text{ A}.$$

In theoretical terms the current takes infinite time to reach the steady state values of V/R and zero, respectively. In practice the values at $t = 5T$ of $0.99 V/R$ and $0.01 V/R$ are likely to be close enough to the steady-state values for approximate analysis.

⟩5.4.3   Predictor–corrector solution of differential equations

The basis of the predictor–corrector method used in the problem-solving programs associated with Chapter 5 is shown in figure A3.

The solution of the differential equation is represented by the increasing exponential curve shown. The technique is based on predicting a new value for the variable, v, from the existing value, using the rate of change. The value to be predicted is V.

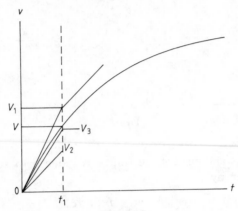

Figure A3

The first prediction is V_1, where $V_1 = st_1$, s being the slope at 0. V_1 is an over prediction which can be corrected in the following manner.

The new value for the slope at t_1 is calculated using the value V_1. If this value of the slope was used from $t = 0$, then the prediction would be V_2, which is an under prediction. A better prediction than either V_1 or V_2 is obtained by taking the average of the two values of slope which produces the prediction V_3.

The process can be used iteratively until two consecutive values are within a specified range.

Variable time intervals
Where the solution of the differential equation is varying less rapidly with time, the same accuracy can be achieved with larger time intervals between successive computations of the variable.

Automatic adjustment can be achieved by doubling the interval until two successive computations are too far apart.

> **5.9   Second-order differential equation responses**

The series $R-L-C$ circuit shown in figure A4 can be analysed in terms of second-order differential equations produced as shown below.

Applying KVL to the circuit gives the equation

$$V = Ri + L\,di/dt + v.$$

Differentiating this equation with respect to time produces

$$dV/dt = R\,di/dt + L\,d^2i/dt^2 + dv/dt$$

Since V is constant, $dV/dt = 0$, additionally $dv/dt = i/C$, hence

$$L\,d^2i/dt^2 + R\,di/dt + i/C = 0$$

$$LC\,d^2v/dt^2 + CR\,dv/dt + v = V.$$

The second-order differential equations can be solved by analytical techniques to produce three different types of functional response depending on the relative magnitudes of the circuit elements.

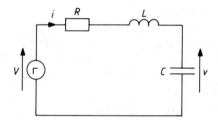

Figure A4

Assuming zero initial conditions for the inductance current and capacitance voltage and using Laplace transforms, the transformed second-order differential equation for current becomes

$$Ls^2\text{LT}[i] - V + Rs\text{LT}[i] + (1/C)\text{LT}[i] = 0$$

where LT$[i]$ means the Laplace transform of i.

Rearranging the equation produces

$$\begin{aligned}\text{LT}[i] &= (V/L)/[s^2 + (R/L)s + 1/(LC)]\\ &= (V/L)/(\{s + [R/(2L)]\}^2 + \{1/(LC) - [R/(2L)]^2\}).\end{aligned}$$

The three functions of time corresponding to LT$[i]$ are as follows:

1. $R < 2(L/C)^{1/2}$

$$\omega^2 = \{1/(LC) - [R/(2L)]^2\}$$

$$\text{LT}[i] = [V/(\omega L)][\omega/(\{s + [R/(2L)]\}^2 + \omega^2)]$$

$$i = [V/(\omega L)]\exp\{-[R/(2L)]t\}\sin(\omega t).$$

2. $R = 2(L/C)^{1/2}$

$$\text{LT}[i] = (V/L)\{s + [R/(2L)]\}^2$$

$$i = (V/L)t \exp\{-[R/(2L)]t\}.$$

3. $R > 2(L/C)^{1/2}$

$$p^2 = \{[R/(2L)]^2 - [1/(LC)]\}$$

$$\text{LT}[i] = [V/(pL)][p/(\{s + [R/(2L)]\}^2 - p^2)]$$

$$i = [V/(pL)\exp\{-[R/(2L)]t\}\sinh(pt).$$

⟩ **6.4   Harmonic products**

The analysis of single-phase and complex-wave power and the process of waveform analysis utilize the sum and difference sine and cosine expressions given below, where m and n are integers corresponding to different harmonics.

1. $\sin(m\omega t)\cos(n\omega t) = 1/2\{\sin[(m + n)\omega t] + \sin[(m - n)\omega t]\}$
2. $\sin(n\omega t)\cos(m\omega t) = 1/2\{\sin[(m + n)\omega t] - \sin[(m - n)\omega t]\}$
3. $\cos(m\omega t)\cos(n\omega t) = 1/2\{\cos[(m + n)\omega t] + \cos[(m - n)\omega t]\}$
4. $\sin(m\omega t)\sin(n\omega t) = 1/2\{\cos[(m - n)\omega t] - \cos[(m + n)\omega t]\}.$

From these expressions it can be seen that the harmonic products produce the sum or difference of other harmonics.

Integral and average values
The integral and average values of the products over one period of the fundamental P for the various possible values of m and n are therefore as follows.

1. $m = n$ or $m \neq n$ integral $= 0$ average value $= 0$
2. $m = n$ or $m \neq n$ integral $= 0$ average value $= 0$
3. $m = n$ integral $= \frac{1}{2}P$ average value $= \frac{1}{2}$
    $m \neq n$ integral $= 0$ average value $= 0$
4. $m = n$ integral $= \frac{1}{2}P$ average value $= \frac{1}{2}$
    $m \neq n$ integral $= 0$ average value $= 0.$

These expressions show that only the products of two sine or two cosine terms with the same frequency have non-zero integral and average values.

⟩ 6.5 Numerical integration

Figure A5 illustrates one method of numerical integration generally referred to as the trapezium rule.

The method consists of finding the area under a curve by summing the areas of a number of strips of equal width, each of which is a trapezium as shown above. If the width of the strips ($t_1 - t_0$), etc, is equal to h, then the integral is approximately given by

Integral =
$$(\tfrac{1}{2}h)[(v_0 + v_1) + (v_1 + v_2) + (v_2 + v_3) + (v_3 + v_4) + (v_4 + v_5) + (v_5 + v_6)]$$

$$= h[(\tfrac{1}{2}v_0) + v_1 + v_2 + v_3 + v_4 + v_5 + (\tfrac{1}{2}v_6)].$$

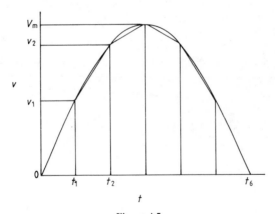

Figure A5

It is clear that the integral becomes progressively more accurate as the strip width is reduced. The process can be used repeatedly, doubling the number of strips after each computation until two consecutive values are within a specified percentage change.

This method of integration is used in the programs with filenames 'INT' and 'HA' for determining the average and RMS values of a sine wave and analysing a complex wave respectively.

⟩ 6.7.5   Solution of equations using interval bisection

The interval bisection method for the solution of equations is demonstrated by using the equation for the operating voltage of the diode in the circuit shown in figure 6.35.

The equation can be expressed in the form

$$\text{FNV}(v) = I_s\ [\exp(v/V) - 1] - (V_s - v)/R$$

where the operating voltage V_0 is determined by FNV $(V_0) = 0$.

It is known from the circuit that the value of V_0 lies between zero and the source voltage V_s. Figure A6 shows a graph of FNV (v) against v, using the example shown in figure 6.35, for values of v from 0 to 0.6 V.

The method of finding V_0 is to approach it by an iterative procedure which continuously bisects the voltage interval in which it is known to exist. The process is continued until the interval is small enough for V_0 to be determined with the required accuracy.

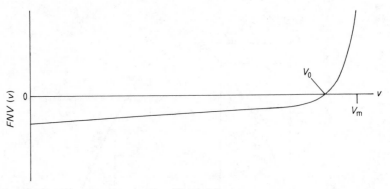

Figure A6

The sign of FNV(v) at the centre of an interval indicates in which half of it V_0 lies. The half interval containing V_0 then becomes the next interval with the previous centre value the next upper or lower limit.

For the example shown:

first interval   0 to V_m $v_1 = \tfrac{1}{2}(0 + V_m)$ FNV$(v_1) < 0$
second interval v_1 to V_m $v_2 = \tfrac{1}{2}(v_1 + V_m)$ FNV$(v_2) < 0$ etc.

If FNV $(v_N) > 0$, then v_N becomes the upper limit of the next interval.

⟩ Appendix B

Software: File Titles, Philosophy and Use

⟩ **B1   Software contents: file titles**

1   Introductory concepts and conventions
1.2	'CURRENT'	Moving charges
	'I'	Current calculation
	'KCL'	Kirchhoff's current law
1.3	'KVL'	Kirchhoff's voltage law
1.5	'VIC'	Voltage and current in passive components
1.6	'CC'	Component combination
1.10	'KLA'	Kirchhoff's laws application

2   Analytical techniques and processes
2.2	'SUPRPOS'	Superposition
2.3	'MESH'	Mesh analysis
	'MA'	Menu for problem-solving programs
2.4	'NODE'	Node analysis
	'NA'	Menu for problem-solving programs
2.6	'NT'	Norton and Thevenin equivalents
2.7	'MAXPOW'	Power transfer
2.10	'PT'	π and T equivalent circuits
2.12	'ATT'	Attenuators

3   Sine waves and basic single-phase theory
3.2	'RMS'	Sine wave squared
	'INT'	Average and RMS values

| 6.8 | 'FDSD' | Frequency-domain software development |
| | 'TDSD' | Time-domain software development |

7 *Practice problems*

| 7.1 | 'SPP' | Solutions to practice problems |
| 7.2 | 'FPP' | Further practice problems |

> **B2** **Software philosophy**

B2.1   General

The basic philosophy of the CIT derives from the concept of an integrated medium consisting of both text and software. This has implications for the presentation of either of the component parts, which are not intended to be entirely freestanding. Some of the text diagrams, for example, only completely achieve their intended clarity when properly viewed on a colour VDU. Similarly with the software, many of the programs contain introductory statements but they are intended to be understood in the context in which they are introduced in the text.

The communication media for the user are the text and VDU. No communication is intended through program statements. The use of the programs does not require them to be modified and does not therefore require programming skill, although knowledge of programming and numerical methods are useful complements.

Broadly the programs fall into two categories: those which produce dynamic displays and others which are designed to solve specific types of problems. For the first category the interaction with the software is minimal but with the problem-solving programs the user is involved in supplying and extracting information to and from the computer.

Application

The programming intention with the problem-solving programs has been to make the interaction process straightforward for the user who has a technical appreciation of the problem. The user should be aware that algorithms will not necessarily work in all circumstances and be conscious of the overall accuracy of the methods used. The programs have been tested with the examples in the text and other similar problems. It is not possible however to test general problem-solving programs for all circumstances. The user must therefore judge the appropriateness of a program for any specific use and the validity of the results produced.

Accuracy
Much of traditional circuit theory consists of exact analysis with ideal models. Because the analysis itself is exact, this can generate a false impression of the validity of the results when related to practical circuits for which the models are only an approximation. With computer methods the analysis itself may be approximate but computing power may allow more realistic models. A real comparison of the two approaches may therefore be between an exact analysis of an approximate model or an approximate analysis of a more realistic one.

A further misconception can arise with accuracy, particularly with electronic calculators, if results are displayed with many digits which are not related to the accuracy of the method of calculation. In most instances the printing format of the programs has been used to reduce the number of digits given in numerical answers. However this is not a substitute for an intelligent interpretation by the user of the validity and accuracy of any problem-solving method used. In broad terms therefore, the user should consider what is required for the solution of a particular problem and then attempt to assess whether the computation meets the requirements.

Other statements on accuracy are made at appropriate places in the text.

B2.2 Computers and electric circuit theory
There are many interesting points to debate on what the impact of the wide availability of individual computing power will be on subjects like electric circuit theory.

At one level the debate will be concerned with the continued relevance of traditional analytical methods which are not required for computer use. These methods are very powerful, have been developed over many years and are an integral part of how many engineers perceive their subject and will not therefore be discarded lightly. These reasons must not however hinder the development of new circuit models which may be more appropriate for the computing tools which become available.

At the educational level, the additional questions which have to be addressed are when in an engineering course students should use industry standard CAD tools and how much should they be aware of the underlying processes whereby those tools produce the answers to specific problems. There is frequently the fear in the mind of the educator and the real danger that the use of computers in problem solving can reduce the activity to a mindless input of data and acceptance of the results.

The CIT would seem to offer a good bridge between fundamental principles and more comprehensive CAD tools. In this instance the algorithms used, while not as powerful or generally applicable as those used in industry standard ECAD software, are likely to be similar in many respects and transparent enough for an understanding of their operation.

Being more specifically applicable the programs can leave the user in closer touch with the problem. Furthermore the data input procedure typically requires the user to be aware of the underlying electrical principles and the computation process. The computer can thus be used to remove the repetitive and lengthy aspects of data manipulation which might normally make some types of problem solving impractical. The interpretation of the results produced can then add another dimension to the understanding of the problem or system being analysed.

〉 B3   Software use

B3.1   Introduction

The individual use of the CIT is based on the personal computer workstation where trends in hardware developments are towards increasing computing speed and memory capacity, higher resolution graphical displays, changing storage media and increased portability, etc, all of which at relatively reduced cost have an impact on the scope and presentation of CIT material. Linked to the hardware developments are parallel developments in the style and presentation of software and one area where this is likely to be most noticeable is in the software–user interface.

Since, at present, text is not as readily modified as computer software it is desirable to present a CIT in a form which is not computer specific in order to take advantage of computer developments and to allow for different software implementations. If this aim is achieved the only apparent differences in relation to the text with different versions of software are likely to be those diagrams which are copies of screen displays; this will be particularly so if different resolution graphics modes are used.

In keeping with this approach the comments here are general with the specific requirements for a particular implementation of the software being left to a software use statement under filename 'SU'.

B3.2   Loading and running the software

Loading
The instructions for loading the software are included with the storage medium in/on which the specific version being used is contained.

Running the programs
The required access to the programs may vary depending on how the CIT is being used, e.g. they may be required randomly for browsing, sequentially as a particular chapter is studied or individually for specific problem solving. Instructions on running the programs are given in the software use statement filename 'SU'.

File structure
The programs are given short titles in the text which relate to their use. The method used for structuring the complete set of program files is included in the software use statement filename 'SU' and for some applications may determine how the programs are accessed.

Program exit
Exit from a program during or after running it, is achieved at two levels; some examples are:

(1)   restarting the program;
(2)   leaving the program and preparing the computer for loading the next one required.

The methods of achieving these exits are likely to be computer specific and are given in the software use statement filename 'SU'. Where appropriate, programs give the user the choice of changing specific parameters prior to recomputing the result.

B3.3   User input

Program progress
For the demonstration type of program the user input is to pace the rate at which the program progresses. This is achieved by the user responding to a prompt to take specific action. This is described for typical operation in the software use statement filename 'SU'. Similarly such instruction

will apply in other programs where pauses are appropriate before continuing to the next stage.

Program options

For a number of programs there are different routes through the program sequence with options such as a component type, variations in graphical display, number of meshes or nodes in a circuit, etc, producing the chosen course. The method of choosing these options may vary depending on the application and is described in the software use statement filename 'SU'.

Data input

The problem-solving programs in particular require the user to enter data, typically the values of circuit parameters. In these instances the value supplied must be followed by the appropriate action to enter it into the program. This is described in the software use statement filename 'SU'.

Note that the problem-solving programs only relate to the models described in the text. They cannot make sense of data which themselves do not make sense in terms of the circuit models given.

〉 Appendix C

〉 Units of Measurement

The information contained in this appendix is extracted from material produced by the National Physical Laboratory and is reproduced with their permission.

Further information: NPL, Teddington, Middlesex, UK.

Quantity, unit, symbol and definition	Realization and use at the National Physical Laboratory
Time: second (s) The second is the duration of 9192 631 770 periods of the radiation corresponding to the transition between the two hyperfine levels of the ground state of the caesium-133 atom.	The second is realized by caesium-beam standards to about 0.1 picosecond—equivalent to a second in 300 000 years. A uniform timescale, synchronized to 0.1 microsecond, is available virtually world-wide by radio transmissions which include broadcasts by satellite.

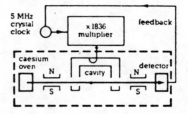

Length: metre (m)

The metre is the length of the path travelled by light in vacuum during a time interval of 1/299 792 458 of a second.

At NPL the metre is realized through the wavelength of the 633 nm radiation from an iodine-stabilized helium–neon laser. The reproducibility is about three parts in 10^{11}, equivalent to measuring the Earth's mean circumference to about 1 mm. The wavelength of this radiation has been accurately related to the new definition.

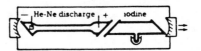

Mass: kilogram (kg)

The kilogram is the unit of mass; it is equal to the mass of the international prototype of the kilogram. This international prototype is made of platinum–iridium and is kept at the International Bureau of Weights and Measures, Sèvres, Paris, France; the British copy (no 18) is kept at NPL.

Kilogram masses and sub-multiples of 1 kg, made from similar materials, may be compared on the NPL precision balance to about a microgram.

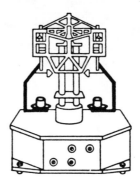

Electric current: ampere (A)
The ampere is that constant current
which, if maintained in two straight
parallel conductors of infinite length,
of negligible circular cross section,
and placed 1 metre apart in vacuum,
would produce between these
conductors a force equal to 2×10^{-7}
newton per metre of length.

The ampere is realized, via the watt,
to about 0.3 μA using the NPL
current-weighing and induced-EMF
method. The ohm is realized at NPL
via a Thompson–Lampard
calculable capacitor to about 0.05 $\mu\Omega$
and maintained via the quantized
Hall resistance to about 0.02 $\mu\Omega$. The
volt is maintained to 0.02 μV using
the Josephson effects of
superconductivity.

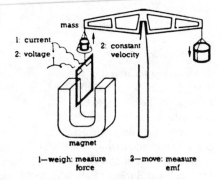

1—weigh: measure   2—move: measure
       force   emf

Thermodynamic temperature: kelvin (K)

The kelvin unit of thermodynamic temperature, is the fraction 1/273.16 of the thermodynamic temperature of the triple point of water. Temperatures may also be expressed in degrees Celsius, where $t(^{\circ}C) = T(K) - 273.16$.

Triple point of water cells are used at NPL to reproduce the triple-point temperature (273.16 K) to 0.1 mK. Other temperatures may be related to this via the International Practical Temperature Scale in terms of which platinum resistance and other thermometers are calibrated within the range of 0.5 to 3000 K.

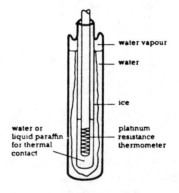

Amount of substance: mole (mol)
The mole is the amount of substance of a system which contains as many elementary entities as there are atoms in 0.012 kilogram of carbon 12. When the mole is used, the elementary entities must be specified and may be atoms, molecules, ions, electrons, other particles, or specified groups of such particles.

The mole is not realized directly from its definition; it can be realized in various indirect ways by use of the concept of 'amount of substance'. The related Avogadro constant (symbol N_A; unit mole^{-1}) is now known to about six parts in 10^7.

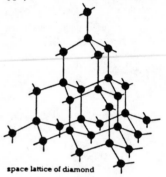

space lattice of diamond

Luminous intensity: candela (cd)
The candela is the luminous intensity, in a given direction, of a source that emits monochromatic radiation of frequency 540×10^{12} hertz and that has a radiant intensity in that direction of 1/683 watt per steradian.

The candela has been realized at NPL with an uncertainty of 0.1% using a cryogenic radiometer which equates the heating effect of optical radiation with that of an electric current. A new solid state photometer has been developed to transfer measurements to light of other frequencies.

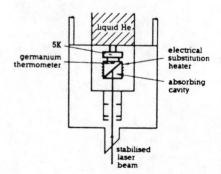

SI supplementary units
Plane angle: radian (rad)
The radian is the plane angle
between two radii of a circle which
cut off on the circumference an arc
equal in length to the radius.

Solid angle: steradian (sr)
The steradian is the solid angle
which, having its vertex in the centre
of a sphere, cuts off an area of the
surface of the sphere equal to that
of a square with sides of length
equal to the radius of the sphere.

These supplementary units are to be
regarded as dimensionless derived
units which may be used or omitted
in the expression for derived units.
NPL angle calibrations achieve an
accuracy of <0.1 arc second. This
would correspond to a navigation
error of about 2 metres in crossing
the Atlantic from New York to
London.

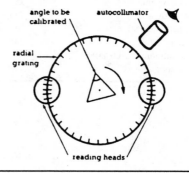

Examples of Si derived units expressed in terms of base units.

Quantity	SI units	
	Name	Symbol
area	square metre	m^2
volume	cubic metre	m^3
speed, velocity	metre per second	$m\,s^{-1}$
acceleration	metre per second squared	$m\,s^{-2}$
wave number	l per metre	m^{-1}
density, mass density	kilogram per cubic metre	$kg\,m^{-3}$
specific volume	cubic metre per kilogram	$m^3\,kg^{-1}$
current density	ampere per square metre	$A\,m^{-2}$
magnetic field strength	ampere per metre	$A\,m^{-1}$
concentration (of amount of substance)	mole per cubic metre	$mol\,m^{-3}$
luminance	candela per square metre	$cd\,m^{-2}$

SI derived units with special names.

Quantity	Name	Symbol	Expression in terms of other units	Expression[a] in terms of SI base units
frequency	hertz	Hz		s^{-1}
force	newton	N		$m\,kg\,s^{-2}$
pressure, stress	pascal	Pa	N/m^2	$m^{-1}\,kg\,s^{-2}$
energy, work, quantity of heat	joule	J	$N\,m$	$m^2\,kg\,s^{-2}$
power, radiant flux	watt	W	J/s	$m^2\,kg\,s^{-3}$
electric charge, quantity of electricity	coulomb	C		$s\,A$
electric potential, potential difference, electromotive force	volt	V	W/A	$m^2\,kg\,s^{-3}\,A^{-1}$
capacitance	farad	F	C/V	$m^{-2}\,kg^{-1}\,s^4\,A^2$
electric resistance	ohm	Ω	V/A	$m^2\,kg\,s^{-3}\,A^{-2}$
electric conductance	siemens	S	A/V	$m^{-2}\,kg^{-1}\,s^3\,A^2$
magnetic flux	weber	Wb	$V\,s$	$m^2\,kg\,s^{-2}\,A^{-1}$
magnetic flux density	tesla	T	Wb/m^2	$kg\,s^{-2}\,A^{-1}$
inductance	henry	H	Wb/A	$m^2\,kg\,s^{-2}\,A^{-2}$
Celsius temperature	degree Celsius	°C		K
luminous flux	lumen	lm		$cd\,sr$[b]
illuminance	lux	lx	lm/m^2	$m^{-2}\,cd\,sr$[b]
activity (of a radionuclide)	becquerel	Bq		s^{-1}
absorbed dose, specific energy imparted, kerma, absorbed dose index	gray	Gy	J/kg	$m^2\,s^{-2}$
dose equivalent, dose equivalent index	sievert	Sv	J/kg	$m^2\,s^{-2}$

a Acceptable forms are for example: $m\cdot kg\cdot s^{-2}$, $m.kg.s^{-2}$, $m\,kg\,s^{-2}$ also m/s, $\frac{m}{s}$ or $m\cdot s^{-1}$.
b See realization of supplementary units.

Examples of SI derived units expressed by means of special names.

| | SI unit | | |
Quantity	Name	Symbol	Expression in terms of SI base units
dynamic viscosity	pascal second	$Pa\,s$	$m^{-1}\,kg\,s^{-1}$
moment of force	newton metre	$N\,m$	$m^2\,kg\,s^{-2}$
surface tension	newton per metre	$N\,m^{-1}$	$kg\,s^{-2}$
heat flux density, irradiance	watt per square metre	$W\,m^{-2}$	$kg\,s^{-3}$
heat capacity, entropy	joule per kelvin	$J\,K^{-1}$	$m^2\,kg\,s^{-2}\,K^{-1}$
specific heat capacity, specific entropy	joule per kilogram kelvin	$J(kg\,K)^{-1}$	$m^2\,s^{-2}\,K^{-1}$
specific energy	joule per kilogram	$J\,kg^{-1}$	$m^2\,s^{-2}$
thermal conductivity	watt per metre kelvin	$W(m\,K)^{-1}$	$m\,kg\,s^{-3}\,K^{-1}$
energy density	joule per cubic metre	$J\,m^{-3}$	$m^{-1}\,kg\,s^{-2}$
electric field strength	volt per metre	$V\,m^{-1}$	$m\,kg\,s^{-3}\,A^{-1}$
electric charge density	coulomb per cubic metre	$C\,m^{-3}$	$m^{-3}\,s\,A$
electric flux density	coulomb per square metre	$C\,m^{-2}$	$m^{-2}\,s\,A$
permittivity	farad per metre	$F\,m^{-1}$	$m^{-3}\,kg^{-1}\,s^4\,A^2$
permeability	henry per metre	$H\,m^{-1}$	$m\,kg\,s^{-2}\,A^{-2}$
molar energy	joule per mole	$J\,mol^{-1}$	$m^2\,kg\,s^{-2}\,mol^{-1}$
molar entropy, molar heat capacity	joule per mole kelvin	$J(mol\,K)^{-1}$	$m^2\,kg\,s^{-2}\,K^{-1}\,mol^{-1}$
exposure (X and γ rays)	coulomb per kilogram	$C\,kg^{-1}$	$kg^{-1}\,s\,A$
absorbed dose rate	gray per second	$Gy\,s^{-1}$	$m^2\,s^{-3}$

Exammples of SI derived units formed by using supplementary units.

Quantity	SI unit Name	Symbol
angular velocity	radian per second	$rad\,s^{-1}$
angular acceleration	radian per second squared	$rad\,s^{-2}$
radiant intensity	watt per steradian	$W\,sr^{-1}$
radiance	watt per square metre steradian	$W\,m^{-2}\,sr^{-1}$

SI prefixes.

Factor	Prefix	Symbol	Factor	Prefix	Symbol
10^{18}	exa	E	10^{-1}	deci	d
10^{15}	peta	P	10^{-2}	centi	c
10^{12}	tera	T	10^{-3}	milli	m
10^{9}	giga	G	10^{-6}	micro	μ
10^{6}	mega	M	10^{-9}	nano	n
10^{3}	kilo	k	10^{-12}	pico	p
10^{2}	hecto	h	10^{-15}	femto	f
10	deca	da	10^{-18}	atto	a

Units, used with the International System, whose values in SI units are obtained experimentally.

Name	Symbol	Approximate value
electronvolt	eV	$1.602\,19 \times 10^{-19}$ J
unified atomic mass unit	u	$1.660\,57 \times 10^{-27}$ kg

〉 Complementary and Further Reading

1. **For numerical methods**

   From the CIT series:

   Harding R D and Quinney D A *A Simple Introduction to Numerical Analysis*
   Harding R D and Quinney D A *A Simple Introduction to Numerical Analysis Volume 2: Interpolation and Approximation*
   Harding R D *Fourier Series and Transforms*

   Other Texts:

   Harding R D *A Mathematical Toolkit* (with software)
   Scraton R E *Basic Numerical Methods*
   Bajpai A C, Calus I M and Fairley J A *Numerical Methods For Engineers and Scientists*
   Hartley P J and Wynn-Evans A *Numerical Mathematics*
   Mason J C *Basic Matrix Methods*
   Mason J C and Stocks D C *Basic Differential Equations*

2. **Electric circuit theory texts covering a similar range of material**

   Fidler J K *Introductory Circuit Theory*
   Edminister J A *Electric Circuits*

3. Further reading in computer methods applied to electric circuit analysis and design

Adby P R *Applied Circuit Theory: Matrix and Computer Methods*
Antognetti P and Massobrio G *Semiconductor Device Modeling with SPICE*
Mastascusa E J *Computer-Assisted Network and System Analysis*

4. General terminology

Young E C *Dictionary of Electronics*

〉 Index